Sustainability in Algerian Higher Education

Salim Bouherar • Sihem Salem

Sustainability in Algerian Higher Education

Balancing Emerging Reforms, Outcomes, and Practices

palgrave
macmillan

Salim Bouherar
Setif 2 University
Setif, Algeria

Sihem Salem
Setif 2 University
Setif, Algeria

ISBN 978-3-031-81346-7 ISBN 978-3-031-81347-4 (eBook)
https://doi.org/10.1007/978-3-031-81347-4

This Palgrave Macmillan imprint is published by the registered company Springer Nature Switzerland AG.
The registered company address is: Gewerbestrasse 11, 6330 Cham, Switzerland

Preface

The intersection of sustainability and education is one of the most pressing and transformative issues of our time. As the global community grapples with the profound challenges posed by climate change, resource depletion, and social inequality, the role of higher education in fostering a culture of sustainability has become increasingly significant. In this evolving landscape, educational institutions are not merely centres of learning but are also crucibles where the values, ethics, and knowledge needed to address the world's most critical issues are forged. This book is an exploration of how these global concerns resonate within the specific context of Algeria, a country rich in history, culture, and natural resources, yet facing its own unique environmental and educational challenges.

Algeria, a nation with a complex socio-political history and a diverse cultural tapestry, is at a pivotal moment in its development. The country's journey from colonisation to independence and its subsequent efforts to build a modern state have been marked by significant achievements as well as enduring challenges. Among these challenges, the imperative is to integrate sustainability into the fabric of Algerian society, particularly within its higher education system. As the world becomes increasingly interconnected, the environmental and social issues that Algeria faces are not isolated phenomena but are intricately linked to global trends and pressures.

In recent years, the Algerian higher education system has begun to recognise the vital importance of integrating sustainability into its curricula, research, and campus operations. This shift is not merely about teaching students the principles of environmental stewardship; it is about equipping the next generation of leaders with the knowledge, skills, and ethical foundation to navigate and address the complexities of a rapidly changing world. Higher education institutions in Algeria are thus tasked with the dual responsibility of advancing knowledge and fostering the critical thinking and problem-solving abilities that students need to contribute to a more sustainable future.

However, as we explore the landscape of higher education in Algeria, we encounter a paradox that is as intriguing as it is concerning. I, Salim, have been fascinated by the perspectives of individuals who approach the current state of Algerian higher education with a purely critical mind. A recent conversation with a colleague brought to light an interesting observation: "the Algerian university system appears to be in a state of plateauing. Whether one invests the maximum effort or only the minimum, the outcomes remain remarkably unchanged", claims the colleague.

This phenomenon suggests that the system may have reached a point where additional efforts, no matter how well-intentioned, do not necessarily translate into better results. In such an environment, the most pragmatic educators are those who provide what they are paid for, meeting the minimum requirements of their roles without striving for further improvement. This is not a critique of individual educators but rather an acknowledgment of a systemic issue that may be stifling innovation and progress. The idea of plateauing is particularly troubling in the context of sustainability, where continuous growth, adaptation, and improvement are essential. If the higher education system remains static, how can it effectively prepare students to tackle the dynamic challenges of sustainability? How can it inspire a generation of thinkers and leaders who are capable of driving the changes needed for a more sustainable future?

The concept of plateauing in education, particularly within the Algerian context, raises important questions about the underlying causes and potential solutions. Is this plateau a reflection of deeper structural issues within the educational system, such as outdated curricula, lack of resources, or insufficient support for educators? Or does it stem from a broader societal context where the value of education, and particularly higher education, is not fully recognised or supported? Understanding the root causes of this phenomenon is crucial for identifying pathways to revitalising the higher education system in Algeria and ensuring that it can play a transformative role in advancing sustainability.

This book seeks to address these questions by providing a comprehensive overview of the current state of sustainability in Algerian higher education. It examines the policies, initiatives, and practices that have been implemented across universities and other educational institutions, while also identifying the gaps and opportunities that remain. Through case studies, interviews, and analyses, we explore how these institutions are working to align their missions with the broader goals of sustainable development, and how they are contributing to the national and global discourse on sustainability.

One of the key themes explored in this book is the concept of education for sustainable development (ESD). ESD is an approach to education that emphasises the integration of sustainability principles across all levels of education, from primary schools to universities. It seeks to empower learners with the knowledge, skills, values, and attitudes needed to contribute to a more sustainable and equitable world. In the context of Algeria, ESD presents both opportunities and challenges. On the one hand, it offers a framework for rethinking and revitalising the higher education system, making it more responsive to the needs of society and the environment. On the other hand, implementing ESD requires significant changes in how education is delivered, assessed, and valued, which can be difficult to achieve in a system that is resistant to change. Algeria's higher education institutions, like those in many other countries, are navigating a complex and often contradictory landscape. On the one hand, there is a growing recognition of the importance of sustainability and the need for education to play a central role in addressing environmental and social challenges. On the other hand, there are significant barriers to achieving this vision, including limited resources, institutional inertia, and competing priorities.

By sharing the experiences and insights gathered in this book, we hope to inspire further dialogue and action, both within Algeria and beyond, on the critical role that education must play in building a sustainable future. This dialogue is not only about identifying the challenges and barriers to sustainability in higher education but also about celebrating the successes and innovations that are already taking place. Across Algeria, there are numerous examples of educators, students, and institutions that are leading the way in integrating sustainability into their work. These examples demonstrate that, despite the challenges, there is tremendous potential for higher education to be a powerful force for positive change.

This book is intended for educators, policymakers, students, and anyone interested in the intersection of sustainability and higher education. It is our hope that it will serve as a valuable resource for those who are working to advance sustainability in Algeria's higher education system and that it will contribute to the ongoing efforts to create a more just, equitable, and sustainable world. In a time of unprecedented environmental and social challenges, the role of education in shaping the future has never been more critical. By exploring the Algerian context, this book aims to shed light on the opportunities and challenges that lie ahead and to contribute to the global conversation on how education can and must evolve to meet the needs of a rapidly changing world.

Setif, Algeria

Salim Bouherar
Sihem Salem

Acknowledgement

We extend our deepest gratitude to everyone who contributed their time, expertise, and insights to this work. To the participants who graciously shared their experiences and knowledge, your input has been invaluable, adding depth and richness to this exploration of sustainability in higher education. Your dedication and willingness to collaborate have been crucial in shaping the narratives and bringing this book to fruition. We are sincerely grateful for your unwavering support and invaluable contributions. A special thanks is owed to Dr. Nadir Boutalbi, whose encouragement and assistance were instrumental in the writing of this book.

Contents

List of Tables

CHAPTER 1

Rationalising the Need for Sustainability Study in Algerian Higher Education

1.1 Background and Context

The Algerian educational system is structured into four distinct stages: primary school, middle school, secondary school, and university. This progression provides a comprehensive educational journey that spans twelve years before students reach higher education (henceforth HE). The journey begins with five years of primary school, followed by four years of middle school, and culminates in three years of secondary school. Children typically start primary school at the age of six, although many begin preschool at age five, depending on their readiness, parental choices, and the availability of spots in primary schools. During the primary school stage, foundational skills are developed, focusing on literacy and numeracy. Pupils are introduced to the basics of the Arabic language, which is the national language of Algeria. Additionally, they start learning French, which holds significant historical and practical importance, and English, which is increasingly emphasised as a global lingua franca. Subjects such as history, geography, and basic sciences are also introduced, laying the groundwork for more advanced studies in middle schools. Middle school builds upon the primary education foundation, deepening pupils' knowledge and skills in various subjects. The curriculum includes more advanced grammar and language studies in Arabic, French, and English. Mathematics and the sciences-physics, biology, and chemistry-become more complex, preparing students for the analytical thinking required in HE. Secondary school, the final stage before university, is crucial for students as they prepare for HE. During these three years, students continue to study a broad range of subjects, but with increasing specialisation and depth. The curriculum is designed to equip students with the necessary knowledge and skills to excel in their chosen fields at the university level. In addition to continued language studies, mathematics, and sciences, students may also study advanced topics in philosophy, literature, and social sciences. Upon completion of secondary school, students take the

S. Bouherar, S. Salem, *Sustainability in Algerian Higher Education*,
https://doi.org/10.1007/978-3-031-81347-4_1

Baccalauréat examination, which determines their eligibility for university admission. This exam is a significant milestone, and its results play a crucial role in shaping students' future academic paths.

Transitioning to university, students are encouraged to choose their major based on their interests, the availability of programmes, and the training capacities of each university. The structure of HE in Algeria differs significantly from the pre-university stages. Universities are composed of various institutions, polytechnics, colleges, and faculties, each specialising in different academic disciplines. These entities operate under the leadership of a local authority, typically a chancellor, who reports to the Ministry of Higher Education. In contrast, the pre-university educational system is governed by local directorates within each Wilaya (province). These directorates report to the national Ministry of Education, ensuring that educational standards and policies are uniformly implemented across the country. This hierarchical structure facilitates a cohesive education system, from primary school through to university. The Algerian HE system is designed to provide specialised training and advanced knowledge in various fields, preparing students for professional careers and contributing to the nation's development. The universities offer a diverse range of programmes and courses, serving the needs of the economy and society. The integration of sustainability practices within these institutions is becoming increasingly important, reflecting global trends and the urgent need for environmentally and socially responsible education.

It is worth mentioning that initiatives towards robust sustainability in HE began to gain momentum after the Al Hirak movement of 2019, when President Abdelmadjid Tebboune assumed power. This political shift brought with it a renewed focus on reforming and improving the HE sector. The government's commitment to fostering sustainable HE practices has become increasingly evident, reflecting a growing awareness of the intricate links between education and economy. Since 2020, there has been a noticeable drive to integrate sustainable economic plans into HE. This effort aligns with the broader goals of economic diversification. One of the significant milestones in this attempt has been the establishment of university business incubators, which support students in developing business models and launching start-up companies. These incubators play a crucial role in bridging the gap between academic knowledge and practical entrepreneurial skills, fostering a culture of innovation and sustainability.

In 2022, two key ministerial decisions revealed the government's commitment to these initiatives. Ministerial Decision 1244 led to the creation of a National Coordinating Committee for Monitoring Innovation and University Business Incubators. This committee is tasked with overseeing the development and implementation of innovative projects within universities, ensuring they align with national sustainability goals. It serves as a central body to coordinate efforts, track progress, and provide strategic guidance on fostering innovation and entrepreneurship in HE. Another significant regulation, Ministerial Decision 1275, also issued in 2022, 27 September, outlines the specific procedures and mechanisms through which students can benefit from establishing start-ups. This decision provides a clear framework for students, detailing the support and resources available to them as they embark on their entrepreneurial journeys. By facilitating the creation of start-ups, the government aims to nurture a new generation of business leaders who are equipped to drive sustainable economic growth. These policy decisions reflect a broader trend within Algerian HE, where there is an increasing emphasis on sustainability and innovation. The political willingness to effect positive change in this sector is evident in the proactive measures being taken to support students and foster a culture of entrepreneurship. An illustrative example of governmental support for innovation and entrepreneurship is Executive Decree No. 20-254, issued on 15 September 2020. This decree mandated the establishment of a national committee tasked with supervising the development and progress of "start-up enterprises", "innovative projects", and "business incubators". The decree outlined the specific roles, composition, and operational mechanisms of this committee, aiming to foster a robust entrepreneurial ecosystem. In a significant move to bolster these efforts, President Abdelmadjid Tebboune established a new ministry dedicated to overseeing the implementation and support of start-up enterprises. This ministry, named the Ministry of Knowledge Economy, Start-ups, and Micro Enterprises, was created to drive the nation's innovation agenda and support nascent businesses. By 1 October 2022, the Ministry of Knowledge Economy, Start-ups, and Micro Enterprises had successfully secured a cooperation contract with the Ministry of Higher Education. This strategic partnership was aimed at promoting economic development and providing robust support for students aspiring to launch their own companies. The collaboration unveils a concerted effort to integrate academic resources with entrepreneurial initiatives, thereby creating a conducive environment for student-led innovation and enterprise.

By linking HE with economic sustainability, the government is laying the groundwork for a more resilient and diversified economy. Moreover, these initiatives highlight the critical role that universities play in national development. By encouraging innovation and providing the necessary infrastructure for start-ups, universities are not only enhancing their educational offerings but also contributing directly to economic development. The focus on sustainability ensures that these contributions are made with a long-term perspective, addressing both current and future challenges.

This book primarily addresses two significant HE reforms introduced in 2019: digitalisation and the implementation of English as a Medium of Instruction (henceforth EMI). The Ministry's efforts to enhance educational quality by replacing paper-based systems with digital technology are commendable, as they contribute to environmental sustainability. However, in the absence of robust infrastructure, the digitalisation policy risks creating disparities, thereby including some while excluding others from the educational experience. Similarly, the adoption of EMI in Algerian HE offers numerous benefits for sustainable development (SD), given English's status as the global lingua franca. Nevertheless, without a comprehensive analysis of the local context and the development of appropriate strategies, enforcing an English-only policy may inadvertently marginalise Algeria's linguistic diversity, which is crucial for achieving a deeper understanding of various disciplines at universities. This book seeks to explore these challenges and provide a nuanced examination of the reforms' potential impact on the Algerian HE system, particularly in relation to the United Nations' (UN) Sustainable Development Goal 4 (SDG4), which aims to ensure inclusive and equitable quality education and promote lifelong learning opportunities for all.

1.2 Rationale for the Book

The UN' 2030 Agenda for SD delineates 17 Sustainable Development Goals (SDGs), with Goal 4 specifically addressing quality education. While reforms in Algeria's HE sector seem to align with these international objectives, it remains uncertain whether the resulting practices and outcomes genuinely reflect a commitment to global sustainability efforts. The importance of sustainability in education is paramount, necessitating a long-term scientific approach and the development of enduring academic initiatives (Ruiz-Mallén & Heras, 2020). The UN's 2030 Agenda aims to enhance the quality of education on a global scale (refer to Chap. 2). Nonetheless, the aspirations and capacities of different countries to achieve this goal can vary significantly, influenced by their unique mechanisms and contexts. The notion of sustainability, particularly concerning quality education, has evolved over decades (Karatsiori, 2023). Consequently, the definition and application of sustainability in HE are shaped by each country's educational ideologies and available resources.

In Algeria, efforts to align with the UN 2030 goals for quality education are evident, particularly following the reforms introduced after 2019. These reforms demonstrate the country's commitment to embedding sustainability within its HE system. However, the measures taken by Algeria to promote sustainability in HE have not been thoroughly or critically explored against UN standards. One possible explanation for this is the region's historically limited engagement in research closely aligned with the economic needs of the Arab world (Hanafi & Arvanitis, 2016).

This book aims to open many doors for embracing more SD studies in HE and to serve as an auditing measure against the standards for implementing the UN goals of quality education. We seek to explore, rather than definitively answer, whether there is a clear trajectory towards achieving sustainability of quality education in Algerian HE (see Chap. 6). Beyond the government's discourse on HE, we aim to transparently examine the concrete steps taken to ensure sustainable HE. In doing so, we aspire to achieve two major objectives: first, to highlight the importance of sustainability in Algerian HE, and second, to emphasise the necessity of transparent reporting on the progress of sustainability initiatives. Particularly, a key objective is to examine the post-2019 government's political will to reform HE for the better. This entails exploring the commitment and actions taken by the Algerian government under President Abdelmadjid Tebboune to enhance sustainability in HE. By examining these efforts, we aim to provide a comprehensive overview of the educational reforms (see Chap. 3). The book emphasises the necessity of transparency and accountability in reporting progress on sustainability initiatives. Promoting open and honest reporting aims to highlight successes, identify challenges, and address potential misconducts. Transparency is crucial for building trust and ensuring that sustainability efforts are credible and impactful.

Furthermore, a fruitful implementation of sustainability in HE has the potential to significantly diversify and strengthen Algeria's economy, which is currently heavily reliant on hydrocarbons. As hydrocarbons contribute approximately 60% of Algeria's gross domestic product and account for 97% of its foreign earnings (Haouas et al., 2024), the economic landscape is vulnerable to fluctuations in global oil and gas markets. By integrating sustainable practices and fostering innovation within HE, Algeria can develop a more resilient and diversified economy. This includes promoting research and development in renewable energy sources, green technologies, and sustainable business practices. Graduates equipped with knowledge and skills in sustainability will be better prepared to lead and participate in new industries, driving economic growth beyond the hydrocarbon sector. Moreover, this shift towards sustainability in HE can attract international partnerships and investments, further boosting economic development. It also aligns with global trends towards greener economies, positioning Algeria as a forward-thinking nation ready to meet the demands of a sustainable future. By investing in education that prioritises sustainability, Algeria would not only enhance its economic prospects but also contribute to global efforts in combating social and environmental issues.

Having said that, the book does not provide a specific model for implementing sustainability in HE. Instead, it focuses on exploring the current status of sustainability in HE by adhering to the UN standards of quality education. This examination is particularly important given the limited discourse on the importance of sustainability within the educational milieu in Algeria. The book aims to bridge this gap by shedding light on the existing state of sustainability practices in Algerian HE institutions. By doing so, it seeks to raise awareness and stimulate discussions on the necessity of integrating SD principles into the educational framework. The exploration considers various aspects such as research initiatives, campus operations, and community engagement to provide a comprehensive understanding of how sustainability is currently being addressed. Through this examination, the book aims to identify strengths and weaknesses in the existing approach, offering insights into areas that require improvement (see Chaps. 4 and 5). By benchmarking against UN standards, it aspires to guide policymakers, educators, and stakeholders in making informed decisions to enhance the role of HE in promoting sustainability. Ultimately, the book serves as a catalyst for change, encouraging a more proactive and strategic approach to sustainability in Algerian HE.

1.3 Significance and Contribution

We approach the topic of sustainability from an educational perspective, highlighting themes where sustainability has the potential to drive economic, social, and environmental improvements. The notable lack of studies on sustainability in Algerian HE has made us deliberate carefully before undertaking this book. Various plans and questions have arisen throughout the process. Should we adopt an insider or outsider perspective in our research? Should we focus on analysing existing models or proposing new ones? How can we effectively capture the nuances of HE reforms with a sustainability focus?

The absence of a substantial theoretical foundation for the Algerian context presented numerous challenges. We felt a profound sense of humility in avoiding heavy assumptions about sustainability in Algerian HE. We recalled a heated discussion among colleagues at the university, sparked by the government's recent interest in implementing digitalisation and introducing EMI in HE. We expressed our frustration, questioning the available resources, timing and necessity of this shift. Many of us have already encountered difficulties in locating digital tools, learning French or mastering academic Arabic. Will digitalisation and adding English to the mix serve as a unifying force or only complicate matters further? We deliberated on the potential educational implications, particularly its impact on local sustainability practices, and decided to approach the topic with careful consideration of its complexities. Thus, we embarked on an exploratory inquiry aimed at providing preliminary insights into whether Algerian HE aligns with the United Nations' SDGs, particularly Goal 4, which focuses on quality education. We believe this study can serve as an initial framework, establishing the foundational elements necessary for future research on the intersection of education and sustainability goals in Algeria.

One of the potential impacts of this book lies in providing insights to researchers working on education for sustainable development (ESD). By filling a significant gap in the literature, this book can serve as a valuable resource for academic researchers who are exploring sustainability within the educational sector. Furthermore, it offers valuable inputs to policymakers by mirroring government actions and examining the current status of sustainability in HE, especially in the post-2019 context. This evaluative aspect is particularly important, as it can inform policy adjustments and help align national educational strategies with global sustainability targets.

Additionally, this book contributes to the field of ESD in Algeria by addressing the scarcity of information and statistics on the topic. By compiling and analysing existing data, and potentially generating new data, this book provides a much-needed foundation for future research and policy development. It highlights the areas where progress has been made and identifies gaps that need to be addressed to achieve ESD outcomes. As such, we acknowledge that substantial progress has been made in enacting the necessary laws and regulations to advance HE but fluctuations in outcomes persist across the sector. This study represents a preliminary step towards understanding the role of sustainability in Algerian HE. Through this work, we aspire to contribute to the ongoing discourse on sustainability in HE, highlighting the potential for transformative change within the Algerian context. Our research not only seeks to lay the groundwork for future investigations but also to inspire a broader dialogue on how educational systems can be reformed to embrace sustainable practices. By emphasising the social, economic, and environmental benefits of sustainability in education, we aim to demonstrate how educationally responsible policies can lead to long-term financial stability and growth. Moreover, this book presents the importance of integrating sustainability into the institutional practices of HE. We argue that sustainability should not be an isolated topic but rather an integral part of the educational experience, preparing students to become leaders and innovators in a sustainable future. This approach ensures that sustainability principles permeate all aspects of academic life, from administration to teaching and research.

1.4 Structure of the Book

This book is structured into six chapters. Chapter 1 serves as an introduction, providing the background and context of the study, particularly focusing on the glotto-political changes that emerged post-2019 in HE. These changes are characterised by a new political will, which appeared to promise significant reforms in HE. The Algerian government has concentrated on two main areas of transformation within HE. Firstly, starting in 2020, the Ministry of Higher Education initiated the digitalisation of the education sector. This process began with e-learning and has since extended to include administrative procedures. Secondly, the government has introduced EMI and established business incubators to support student-led business projects within universities. These initiatives reflect the government's intention to enhance the quality of education and align it with the job market, thereby promoting sustainability. This chapter also addresses the importance of incorporating sustainability issues into Algerian HE. It lays the groundwork for future empirical studies that will explore the role of sustainability in HE in greater detail.

Chapter 2 investigates the intersection of sustainability and HE in Algeria, focusing on the implementation of the United Nations' SDGs within the context of Algerian universities. The chapter examines how we approach the term sustainability in light of the nuanced concepts of its applications. It discusses sustainability in terms of meeting the needs of the current generation without compromising the ability of future generations to meet their own needs. This chapter also serves as a reference point for the discussions in Chaps. 4 and 5. Despite global calls for action, Algeria faces unique challenges in aligning its educational practices with sustainability objectives, hindered by socio-economic and political factors. It also explores the influence of government policies and centralised decision-making on the progress towards ESD. Additionally, the chapter explores innovative approaches from Gulf countries, highlighting successful models that Algeria can adapt to enhance its HE system. Ultimately, this analysis highlights the need for a contextualised approach that addresses Algeria's specific needs and challenges, aiming to build a resilient and sustainable HE framework.

Chapter 3 discusses the diverse reforms implemented by the Ministry of Higher Education under successive Algerian governments, examining how political ideologies have shaped the Algerian university system since independence in 1962. Key reforms, such as the critical 1971 HE reform following the nationalisation of oil and gas resources, aimed to assert Algeria's autonomy by diminishing French influences. These reforms reflect the broader socio-economic objectives of the administrations in power and highlight the dynamic interplay between national policy, resource management, and educational restructuring. The chapter also discusses the abrupt shift from French to Arabic in HE, a move critiqued by some scholars for its lack of comprehensive linguistic analysis, yet acknowledged for addressing local labour demands and boosting the economy. Institutions like the University of Science and Technology-Houari Boumediene (USTHB) played a pivotal role in producing highly skilled graduates to meet industrial needs. Subsequent sections explore the evolving educational landscape under different administrations, including the impact of President Chadli Bendjedid's reforms and the influence of the Berber and Islamic movements. The chapter culminates with an analysis of the significant 2004 LMD (Licence, Master, Doctorat) reforms and the ongoing glotto-political dynamics, particularly the post-2019 shift towards EMI, reflecting a continued effort to align HE with national aspirations and global standards.

Chapter 4 examines the significant milestone announced by the Minister of Higher Education regarding the digital transformation of university. The chapter discusses this strategic shift to address logistical challenges associated with physical university attendance and reduce the burdens of traditional paperwork. This chapter also highlights the challenges that persist, particularly with the digitalisation reliability, such as the instability of the PROGRES platform used for student and teacher evaluations. Furthermore, this chapter critically examines the implications of digitalisation in HE, focusing on its impact on students and educators and its alignment with SD outcomes. Through a survey conducted across 47 universities in Algeria, this chapter explores the experiences of teachers and students with e-learning and digitalisation. The investigation reveals centralised and inflexible plans for digitalisation while plans to digitalise HE seem to be more aligned with administrative procedures than academic ones. The findings suggest that while digitalisation has the potential to enhance efficiency and transparency, some challenges remain that must be addressed to realise these benefits fully.

Chapter 5 explores the implementation of EMI in Algerian HE, a strategic initiative by the government aimed at bolstering the international standing of Algerian universities. This chapter critically examines the multifaceted challenges and pivotal factors influencing the successful deployment of EMI. Central to this analysis are the government's strategic frameworks and their impact on teacher training, curriculum development, and policy formulation. The role of local and national committees in overseeing EMI implementation is explored, highlighting both their advisory capacity and the constraints they face in ensuring compliance. Drawing on data from 31 participants, including teachers and students from various universities, the chapter identifies critical barriers such as insufficient English teaching resources, linguistic marginalisation, and the complexities of transitioning to an English-based curriculum. Moreover, it critiques the predominant focus on educator training at the potential expense of enhancing student English proficiency. The chapter concludes with a call for a nuanced multilingual approach to enriching Algerian HE.

Chapter 6 summarises the key findings and offers recommendations for future research and action, addressing the complexities and challenges of integrating sustainability into Algerian HE. It highlights three primary findings: the practices of digitalisation, the implementation of EMI, and the lack of an integrated approach across all aspects of HE. The chapter recommends conducting more inclusive studies to investigate the outcomes of SDG4 on the local economy. It also suggests involving a larger participant number in future studies to enhance the credibility of the findings. For decision-makers, the chapter offers several action-oriented recommendations. Addressing the ongoing issue of insufficient technological infrastructure is critical for the government. Additionally, the EMI approach should be introduced gradually to avoid marginalising speakers of other languages. Finally, policies related to SD need to be taken seriously and should aim for long-term planning.

1.5 Conclusion

We acknowledge that the topic of sustainability has not been widely and fully discussed in this book due to the scarcity of resources, especially in Algerian HE. However, through the various chapters of the book, we have explored how this concept resonates with the educational, economic, and governmental practices in the Algerian university system. While this may not provide a comprehensive analysis of sustainability, it opens doors for future research that can focus on specific aspects of sustainability in HE, positioning this book as an overview and starting point for more detailed investigations. The book provides various implications for policymakers and researchers in the field, particularly discussed in Chap. 6. Although sustainability is a relatively new topic in the context of HE, there are areas of improvement highlighted throughout the book. One of the central points of the book is that the Algerian government has shown significant interest in HE as a crucial factor in SD. However, it is the mechanisms implemented to achieve this goal that are the subject of our investigation, such as digitalisation, EMI, and university business incubators.

Therefore, we pose the question: should we scrutinise the Algerian government's intentions for improving sustainability plans, or should we focus on examining how these intentions are implemented? As this book elucidates, it is primarily the implementations that we explore. President Tebboune's administration has demonstrated a strong will to rectify many of the previous government's misconducts pre-2019. Thus, this book serves as a critical document examining the efforts and shortcomings in pursuing sustainability in HE. It is clear that the journey towards sustainability in Algerian HE is ongoing. This book highlights the initial steps and identifies the significant challenges and opportunities that lie ahead. It serves as a call to action for scholars, policymakers, and practitioners to engage in more detailed and targeted research, essentially contributing to a more sustainable and resilient HE system in Algeria. By shedding light on the current practices and offering critical insights, this book aims to inspire a more profound commitment to sustainability in HE. It encourages continuous dialogue and collaboration among all stakeholders to create a sustainable future for Algerian universities and beyond.

In addition, we suggest that future research should explore the intersection of sustainability with other pressing issues in HE, such as equity, access, and internationalisation. Understanding how these dimensions interact can provide a more holistic approach to achieving sustainability goals. This interdisciplinary perspective is crucial for developing innovative solutions that address the complex challenges facing HE today. Ultimately, this book serves as both a critique and a roadmap, guiding stakeholders towards a more sustainable future in Algerian HE. The insights and recommendations offered herein are intended to spark further inquiry, dialogue, and action, contributing to the ongoing efforts to create a more equitable and inclusive ESD.

References

Hanafi, S., & Arvanitis, R. (2016). *Knowledge production in the Arab World: The impossible promise*. Routledge.

Haouas, A., Ochi, A., & Labidi, M. A. (2024). Sources of Algeria's economic growth, 1979–2019: Augmented growth accounting framework and growth regression method. *Regional Science Policy & Practice, 16*(3), 1–19.

Karatsiori, M. (2023). In the pursuit of "Quality Education": From ancient times to the digital era, can there be a consensus? *Cogent Education, 10*(2), 1–33.

Ruiz-Mallén, I., & Heras, M. (2020). What sustainability? Higher education institutions' pathways to reach the agenda 2030 goals. *Sustainability, 12*(4), 1–18.

CHAPTER 2

Unveiling Sustainability: Exploring Concepts, Goals, and Hurdles

2.1 Introduction

The concept of sustainability was introduced by the United Nations (UN) as a global imperative aimed at reshaping prevailing world conditions. The sustainable development goals (SDGs) serve as a collective call to action for all nations, urging them to achieve significant milestones by 2030. However, the atrocities of the educational spectrum in Algeria surpassed its ability to maintain stable sustainability routes to future generations. Despite the commitment of higher education institutions (henceforth HEIs) to contribute to sustainability (Webb & Rodriguez, 2022), the Algerian context seems to offer unique experiences in the pursuit of the SDGs. The specific circumstances in Algeria present a complex interplay of factors that demand careful considerations when aligning educational practices with global sustainability objectives. Debunking the arguments supporting Algeria's adherence to a sustainable path towards the UN's SDGs would not necessarily be a challenging task. One notable aspect is the need for a contextualised approach that acknowledges the socio-economic landscape of Algeria. Tailoring sustainability initiatives to align with the country's specific needs and challenges is of paramount importance. The higher educational system must not only impart knowledge but also address the socio-economic disparities that may impede the effective implementation of sustainable practices. What is deemed a difficult task is to build a sustainable model that aligns with the current socio-political and economic dynamics in Algeria.

One of the questions confronting us in this chapter is how to conceptualise sustainability in Algerian higher education (HE). Can we synchronise sustainability goals with outdoor education to advance system thinkers (Karaarslan-Semiz, 2022) as a model for Algerian HE challenges? Can a reformed religion-based HE (Assalihee & Boonsuk, 2022) present an alternative to the current realities of Algerian HE? Perhaps the development of a sophisticated model for monitoring, evaluation, accountability, and learning (Gichuhi et al.,

S. Bouherar, S. Salem, *Sustainability in Algerian Higher Education*,
https://doi.org/10.1007/978-3-031-81347-4_2

2022) can scrutinise issues and propose solutions to educational challenges. Every model is distinctive to a specific educational context, but what is Algeria's fitting model? An answer to that question necessitates an exploration of HE realities that impede sustainability. Issues such as the centralisation of decision-making and politicised reforms significantly hinder progress. The concentration of decision-making power in centralised structures often limits flexibility and responsiveness to diverse needs. Additionally, politically motivated reforms may prioritise short-term gains over long-term sustainability, creating challenges for the development of a robust and adaptable HE system.

This chapter builds on the UN's SDG guidelines on establishing sustainability in education against the ongoing practices in the Algerian university. The fourth goal of the UN's SDGs is quality education, which is the main endeavour of the chapter. In this chapter, we explore how Algerian universities are incorporating the principles of sustainability into their curricula, research initiatives, and operational practices. We also examine the extent to which these universities are promoting critical thinking, creativity, and problem-solving skills among their students, preparing them to tackle the complex challenges of a sustainable future. The chapter also examines the effectiveness of Algerian universities in fostering collaboration and partnerships with stakeholders from various sectors, including government, industry, and civil society, to advance SDGs. Additionally, it discusses the challenges and opportunities faced by Algerian universities in their pursuit of sustainability in education.

Algerian universities are heavily influenced by the government's political agenda due to their reliance on government funding (Bouherar & Ghafsi, 2021). This chapter explores the political initiatives aimed at enhancing the visibility and ranking of Algerian HE on the global stage. Specifically, it discusses the role of the government, represented by the Ministry of Higher Education, in enacting relevant policies and decrees to elevate the quality of education. As an illustration, between 22 September and 17 November 2022, eight decisions were issued to improve the quality of education, addressing various aspects such as digitalisation, ranking, and promoting the university's contribution to the economy (MESRS, 2023b). These efforts underscore the government's commitment to strengthening Algerian HE and positioning it as a key driver of national development. However, realities can suggest different discourse of the government's actual statements. The chapter also discusses the challenges that hinder the establishment of a sustainable HE in Algeria. We maintain that ensuring the sustainability of HE in Algeria remains a complex challenge. One of the primary concerns is the persistent financial constraints faced by Algerian universities. The overreliance on government funding limits the ability of universities to invest in infrastructure, recruit and retain qualified faculty, and provide adequate resources for students.

The chapter culminates by examining exemplary case studies from Gulf countries that have pioneered innovative approaches to Education for Sustainable Development (ESD). Nations like Qatar, the United Arab Emirates, and Saudi Arabia have implemented distinctive strategies to elevate HE and foster quality learning. These countries have recognised the pivotal role of education in driving national development and have embarked on ambitious endeavours to enhance their HE systems. Qatar (see Cochrane & Al-Hababi, 2023), for instance, has established Education City, a vibrant hub of knowledge and innovation that brings together world-class universities, research institutions, and technology companies. This initiative has attracted renowned scholars and fostered a dynamic environment for collaboration and knowledge exchange. Similarly, the United Arab Emirates (see Alfaki & Ahmed, 2016) has made significant investments in upgrading its universities and attracting top-tier faculty. The country has also prioritised the development of specialised research centres in areas such as renewable energy, nanotechnology, and artificial intelligence. These efforts have propelled the UAE's HE system onto the global stage. The success of these Gulf countries in promoting ESD offers valuable lessons for Algeria. By emulating their commitment to innovation, collaboration, and quality, Algeria can effectively address the challenges it faces and build a thriving HE system that prepares its graduates for the demands of the twenty-first century.

2.2 Conceptualising Sustainability: Navigating Complexity for a Better Present and Future

The term "sustainability", though phonetically straightforward, proves elusive in its conceptualisation. It denotes the enduring maintenance and availability of resources over time. Sustainability implies the capacity for prolonged maintenance. A society deemed unsustainable is incapable of enduring over an extended period and is destined to cease functioning at some juncture (Heinberg & Lerch, 2010). Sustainability embodies complexity due to its rapid and extensive adoption in delineating diverse contemporary issues. Since its inception in the Brundtland Report of 1987, the concept of sustainability has evolved into a seminal subject for scholarly inquiry within various fields, including business (Bader, 2005; Lüdeke-Freund et al., 2018; Sharma et al., 2008), economy (Ayres et al., 2007; Christensen et al., 2021; Elliott, 2005), environment (Brinkmann, 2020; Ekins, 2011; Schileo & Grancini, 2021), politics (Birnbacher & Thorseth, 2015; Meuleman, 2013; Scoones, 2016), society (Davidson, 2010; Dempsey et al., 2011; Larimian & Sadeghi, 2021), and education (Dale & Newman, 2005; Ruiz-Mallén & Heras, 2020; Sterling, 2001). Whilst sustainability traditionally focused primarily on environmental concerns, it has transformed into a comprehensive term that embraces a wide range of dimensions, encompassing business, economics, politics, society, and education. New conceptions of sustainability, based on a spectrum of critical issues, have been explored in empirical studies. These include environmental sustainability (e.g. Goodland, 1995), business sustainability (e.g. Bansal & DesJardine, 2014), economic sustainability (e.g. Ikerd, 2012), political sustainability (e.g. Patashnik, 2003), social sustainability (e.g. De Fine Licht & Folland, 2019), and educational sustainability (e.g. Alismaiel, 2021). Certain scholarly inquiries approached sustainability as an overarching concept and explored its implementation by considering its three fundamental pillars: the environment, the economy, and society (Bandarage, 2013; Giddings et al., 2002; Li & Yi, 2020). Research approaches to sustainability vary, highlighting its status as a concept open to ongoing debate and discussion.

Concurrently, sustainability has captured the interest and consideration of policymakers engaged in educational initiatives. The integration of sustainability into the educational sphere has given rise to diverse concepts, such as "sustainability education" (SE), "education for sustainability" (EfS), "education for sustainable development" (ESD), and "sustainable education". These notions are commonly employed interchangeably and regarded synonymous, as indicated in the literature. However, despite the presence of similar ideas, these terms may not inherently convey identical meanings (Jeronen, 2022). Sustainability education, according to Moore (2005, p. 78), is "interdisciplinary, collaborative, experiential, and potentially transformative". In the context of HE, a curriculum that integrates experiential learning in an interdisciplinary collaboration can foster critical and creative thinking among students. This approach has the potential to transform how they address complex issues to navigate the rapidly changing world. Moore (2005, p. 78) defines sustainability as "a concept, a goal, and a strategy". She goes on to elaborate that sustainability education is "a process of creating space for inquiry, dialogue, reflection, and action about the concepts and goals of sustainability" (ibid.). From her interpretation, sustainability education involves addressing sustainability as a concept by reconciling ecological integrity, social justice, and the well-being of all living systems on earth. As a goal, it strives to establish a world that is both ecologically and socially just, all within the limits set by nature, safeguarding the interests of future generations. Additionally, it serves as a strategy to move towards a sustainable future through the creation of a space for dialogue, inquiry, reflection, and action. In essence, sustainability education entails integrating these principles of sustainability into curriculum content and teaching practices, with an emphasis on transformative learning utilising interdisciplinary collaboration and experiential approaches to address the present challenges and attain a future that is environmentally and socially sustainable.

It is crucial to recognise that the incorporation of sustainability into educational programmes encompasses diverse terminologies. The interpretation, policy formulation, and implementation may vary depending on the chosen terminology. Some educators prefer the term education for sustainability (EfS) and have included it in their HE programmes, such as London South Bank University. Wade (2008) argues their team selected EfS over other conceptions because it is a versatile framework that interprets sustainability across different scales, ranging from personal to global. It encourages exploration of educational methods and purposes while promoting open commitment to values that drive sustainable change in individuals and society. EfS aims to facilitate a holistic understanding of sustainability and foster positive transformations.

Wade (2008, p. 32) considers EfS to be "possessing both the *process towards* and the *vision of* sustainability". This field promotes a process of interdisciplinary learning by enabling different disciplines to interpret EfS using their own visions, contributing to a thorough understanding of sustainability. However, Sterling (2001) asserts that there is a deficiency in clearly defining the required vision and strategically understanding the process of advancing towards that vision, considering the influence of the prevailing social and educational programmes. He believes that effective change, at any level, necessitates a harmonious interplay and reciprocal influence between the visionary and the practical aspects. Sterling (2001) contends EfS remains prone to adaptation by the mainstream. Wade (2008) also acknowledges that the realm of EfS has generated innovative ideas and concepts, necessitating new vocabulary and potentially demanding new language. Given the contentious nature of sustainability, its incorporation into HE is susceptible to various debatable conceptualisations.

The term EfS was intentionally chosen to foster an inclusive stance that embraces different viewpoints and interpretations of sustainability. It was also selected to critically examine the predominant Western-centric discourse on development, as encapsulated in the concept of ESD (Wade, 2008). However, the UN has designated ESD not as solely a Western responsibility, but rather as a global imperative. Despite extensive research on the subject being prevalent in Western countries, as evidenced by scholarly work, it is essential to explore ESD from non-Western perspectives. Educational establishments, regardless of their types and levels, play a crucial role in the worldwide mission of steering humanity towards sustainable development (SD). Thus, countries need to reform their educational policies to effectively incorporate ESD into their curricula, teacher training programmes, as well as teaching and learning materials and environments. If a country enforces a robust policy requiring ESD but lacks the corresponding support in terms of ESD-focused teacher training and the creation of learning materials aligned with ESD principles, there will be a notable deficiency in the execution of the policy (Didham, 2018). The HE sector, for instance, must have the essential resources and capacities to proficiently implement and deliver ESD. This implies the necessity of balancing reform initiatives and practical implementations to enhance overall outcomes. Attaining equilibrium requires a dedicated focus on the following elements: (1) strengthening governance structures, accountability measures, and collaborative partnerships; (2) implementing effective coordination strategies; (3) establishing rigorous monitoring, reporting and evaluation frameworks to inform evidence-based policies; and (4) prudently allocating financial resources (UNESCO, 2016). By enacting these cornerstones, ESD cultivates well-informed decision-making and conscientious actions to uphold environmental integrity, ensure economic viability, and promote social justice across generations, while honouring cultural diversity.

ESD enhances the quality and relevance of education. It improves cognitive, socio-emotional, and behavioural aspects of learning, constituting a lifelong learning process. ESD comprises content, outcomes, pedagogy, and the learning environment to create a holistic and transformational educational experience (UNESCO, 2023). The content of ESD focuses on addressing interconnected global challenges, including inequality, climate change, unsustainable resource use, loss of biodiversity, and economic issues. The outcomes of investigating these intricate issues encompass developing responsible global citizens, who, through informed decision-making, critical actions, and collaboration, contribute to achieving sustainability across social, cultural, environmental, and economic dimensions in the world. The pedagogy employed to explore these global issues is a critical approach that promotes a learner-centred teaching and learning process, empowering individuals to engage in problem-posing and take sustainable actions. The learning environment in ESD is interactive and collaborative, providing essential resources for a comprehensive exploration of complex issues. It fosters transformative learning, empowering individuals to address present challenges and contribute to solutions for current and future generations. UNESCO's (2023) efforts in the field of ESD revolve around five key areas: (1) advancing policy; (2) transforming learning environments; (3) building capacities of educators; (4) empowering and mobilising youth; and (5) accelerating local level action. Countries actively engaged with UNESCO in advancing ESD are integrating it into their curriculum development, education planning, and policy frameworks. They are committed to establishing a holistic learning environment that embraces technology and experiential learning for a profound understanding of environmental, social, and economic interdependencies. They are also dedicated to empowering their educators by providing training and support. Furthermore, these nations are directing their efforts towards engaging youth, promoting activism, fostering leadership development, and involving young people in the decision-making process. Local initiatives are implemented through projects aimed at addressing global challenges. These endeavours collectively create a comprehensive educational framework that promotes a deep understanding of SD and instil a sense of responsibility for building a more sustainable and equitable world.

Substantial progress has been attained through the fields of sustainability education, EfS, and ESD. Nonetheless, despite these commendable strides, a discernible deficit persists in effectively aligning attitudes, values, and behaviours with the imperative adoption of sustainable lifestyle (Jeronen, 2022; Sterling, 2001). In response to the pressing issue of unsustainability, a novel conceptual framework emerges. Sterling (2001) proposes the term sustainable education (SE) as a shift from "education for something" to education in its entirety. This encompasses its paradigms, policies, purposes, and practices—all interconnected—and evaluates its suitability for the current era. He contends most educational institutions perpetuate unsustainable values and practices, emphasising competition and consumption rather than fostering care and conservation. Formal education is thereby implicated in the predicament of unsustainability. For this reason, Sterling (2001, p. 14) considers SE as:

> A change of educational culture which both develops and embodies the theory and practice of sustainability in way which is critically aware. This would be a transformative paradigm that values, sustains and realises human potential *in relation to* the need to attain and sustain social, economic, and ecological wellbeing, recognising that they are deeply interdependent.

Promoting SE within HE requires a departure from traditional educational norms. It involves actively advancing and incorporating both the theory and practice of sustainability with a critical awareness. This approach represents a transformative paradigm that emphasises the appreciation, preservation, and actualisation of human potential within the broader framework of fostering and maintaining the interconnected well-being of society, economy, and ecology. This form of education acknowledges the merits of past ideas and practices, simultaneously ensuring the preservation of future progress. SE entails a fundamental paradigmatic transformation, while ESD adheres to the prevailing paradigm. The impediment to enacting the adaptation and reorientation of education towards sustainability is, in part, attributable to a deficiency in envisioning and articulating the foundational principles underpinning such a reorientation (Sterling, 2001). However, ESD involves more than just the integration of sustainability principles into existing educational paradigms. It also encompasses the reform of educational policies; the transformation of learning environments; the provision of support and resources for educators; the recognition of the interplay between society, culture, environment, and economy; and a focus on promoting sustainable lifestyle. The question of whether Western and non-Western educational systems are effectively promoting sustainability prompts not a shift away from ESD, which inherently emphasises educational considerations, but rather necessitates scrutiny of whether sustainability concerns are adequately addressed and whether there exists a harmonious balance between reform initiatives and desired outcomes. These issues are explored within the context of Algeria in the present study.

ESD has attained global recognition as a consequence of the initiatives undertaken by the UN. This has prompted a paradigm shift in policies and the instigation of new reforms on an international scale. Despite the global commitment to ESD arising from the UN decade of ESD (2005–2014), which aims to integrate the principles and practices of SD into all facets of education and learning for the purpose of fostering a more sustainable and equitable society, there's no distinct branch explicitly designated as ESD or related concepts within the Algerian educational framework. However, there may be an evolution of the education system towards SD through the implementation of new policies. The lack of a specific emphasis on ESD within the curriculum could account for the scarcity of empirical studies examining the integration of ESD in Algerian HE. ESD serves as a valuable framework for advancing high-quality education and ensuring equitable access, particularly in the post-COVID-19 landscape. Educational policies assume a central role in facilitating the effective enactment of ESD, delineating its influence on and benefits for curriculum design, teacher training, the creation of educational materials, and the learning milieu. ESD offers a comprehensive framework for building a more sustainable and just future. In the wake of the pandemic's repercussions, the adoption of ESD becomes not only a strategic decision but an imperative to cultivate responsible global citizens capable of contributing to the well-being of the planet and its inhabitants.

The twenty-first-century global pandemic has profoundly influenced various aspects of the world in the aftermath of COVID-19, primarily by presenting multifaceted challenges in healthcare, economy, politics, and education. This has necessitated crucial adjustments across different facets of life, particularly within the HE system. New reforms have been introduced and implemented to ensure preparedness for such circumstances. Significant emphasis has been placed on digitalisation, which served as a recourse during the COVID-19 pandemic, aiding in the continuity of education during challenging times. Algeria is one of the nations that have acknowledged the numerous benefits of digitalisation in HE, such as facilitating administrative work and providing access to education at any time. Consequently, the "zero-paper" policy was heralded at the commencement of the academic year 2023. Digitalisation enhances the accessibility and visibility of HE globally, and English, being a widely used language worldwide, plays a pivotal role in this context. The shift from French to English as a medium of instruction (EMI) is another reform that aligns with digitalisation. It is imperative to scrutinise whether these new reforms demonstrate sustainability and align with the UN's SDGs, particularly Goal 4, which underscores the importance of equality and inclusiveness in education.

2.3 Sustainable Development Goals: A Global Imperative for Positive Change

One of the paramount concerns of our era is the global imperative to effect positive change by preserving our planet's resources while simultaneously fostering economic prosperity and societal welfare for a growing population. This significant endeavour is encapsulated within the framework of SD. Essentially, SD addresses the current socio-economic needs without depleting resources or causing environmental harm that could impede future generations from fulfilling their own requirements. It underscores responsible, long-term planning to safeguard the well-being of both present and future generations. As articulated by the Brundtland Report of 1987, SD represents the form of development that "meets the needs of the present generation without compromising the ability of future generations to meet their own needs" (p. 16). Despite the widespread adoption of this definition, it has encountered criticism for its failure to explicitly recognise the unsustainable utilisation of non-renewable resources and for overlooking the rise in population (see Heinberg & Lerch, 2010). However, the report acknowledges that SD is restricted by existing technology and social structures, but it suggests that both can be controlled and enhanced to facilitate a new phase of economic growth. Rather than portraying SD as a static state of harmony, the report presents it as a dynamic process of change wherein the utilisation of resources, the direction of investments, the focus of technological progress, and institutional changes are harmonised with the needs of both present and future generations. The report emphasises that addressing the obstacles of SD requires adapting and evolving the approaches we employ to respond to the changing circumstances.

Elucidating the concept of SD is a challenging task, much like the complexities associated with sustainability. The continuous evolution of SD poses difficulties for scholars, hindering precise understanding and articulation of its exact meaning. It is commonly perceived to involve three interrelated components: the environment, society, and economy (Boeve-de Pauw et al., 2015; McKeown et al., 2002). Initiated by the UN, countries have agreed to pursue SD. The nexus between education and SD is intricate, yet education plays a pivotal role in a nation's capacity to progress and attain sustainability. Educational forums are central in advancing SD through active participation in ESD. This involvement focuses on comprehensive educational initiatives that address environmental, social, and economic concerns. The interconnectedness of these dimensions underscores the necessity for a holistic strategy to achieve a sustainable lifestyle. An illustrative example of this endeavour is the transition from traditional paper-based materials to digital tools in education. This shift not only contributes to environmental preservation by reducing paper consumption but also enhances students' accessibility to education, fostering social inclusivity across diverse locations and times. Moreover, the adoption of digital tools has the potential to stimulate economic growth across various industries by fostering technological innovation and enhancing operational efficiency. These initiatives exemplify the multifaceted impact that educational reforms can have on promoting sustainability. By addressing environmental, social, and economic aspects, education becomes a powerful tool in shaping a more sustainable future.

The UN established SD for countries to ensure the well-being, equity, and environmental responsibility of future generations. This involves addressing present needs while fostering inclusive policies. SD is essential in contemporary communities to instigate positive transformations. Ongoing adverse effects of human activities on the environment, such as climate change, desertification, pollution, loss of diversity, and depletion of aquifers, coupled with social issues like gender inequalities, discrimination, and unequal access to education and healthcare, as well as economic challenges such as poverty, insufficient nutrition, limited access to potable water, and the prevalence of infectious diseases, highlight the urgent need for SD. These challenges emphasise the vital role of SD in driving positive environmental, societal, and economic changes in both affluent and impoverished nations. HEIs can be a powerful force for SD. Their mission involves improving the quality and effectiveness of education through embracing innovative practices, conducting impactful research, and implementing policies that not only transform the current landscape but also foster a future marked by inclusivity, justice, and ecological vitality. In this way, HEIs have the potential to nurture individuals who actively contribute to SD by integrating sustainable practices across various dimensions of their structures, including educational framework, research pursuits, daily operational activities within their campuses, community outreach initiatives, and the meticulous assessment and transparent reporting of their sustainable efforts (Aleixo et al., 2016; Fischer et al., 2015; Lozano et al., 2015). More HEIs worldwide are embracing the commitment to SD by signing declarations or charters, or participating in SD initiatives to integrate sustainable practices into their operations.

However, some institutions have not officially adopted a declaration or charter, yet they are actively involved in incorporating SD initiatives across their campuses (Lozano & Watson, 2013). Meanwhile, others have formally endorsed a declaration, a charter, or an SD initiative but have failed to implement it effectively within their systems (Lozano et al., 2015). The integration of SD into HEI systems does not guarantee genuine commitment to SD in their daily practices. It is incumbent upon HEIs to assume the responsibility for implementing SD, necessitating the establishment of rigorous procedures to monitor and assess their progress in this regard (Walton, 2000). European HEIs are at the forefront of actively embedding SD into their systems (Disterheft et al., 2012; Karatzoglou, 2013; Lozano et al., 2015). A limited number of studies have explored the incorporation of SD within HEIs in the Arab context, encompassing countries like Algeria. This scarcity may be attributed to the relatively low prevalence of charters, declarations, or SD initiatives within the region. Notably, in Algeria, these themes have been recurrent in official discourse. The recent wave of reforms, including the incorporation of EMI and the digitalisation of HEIs, appears to represent a strategic approach aimed at advancing SDGs. In this book, our primary objective is to investigate the extent to which these policies, along with associated practices and outcomes, align with the fundamental principles of SD. This inquiry aims to illuminate the efficacy of the aforementioned initiatives in fostering sustainability within the educational landscape of Algeria, ultimately contributing to the broader discourse on the intersection of HE and SD in the Arab context and on a global scale.

Leaders from around the world convened in 2015, forging a groundbreaking commitment to ensure the rights and well-being of every individual on a thriving, healthy planet. This transformative effort gained momentum with the adoption of the 2030 agenda for SD and its 17 SDGs. These global goals were embraced by the UN, serving as a universal call to eliminate inequality and ensure that all people experience peace and prosperity by the year 2030. The imperative for action embedded in the SDGs is applicable globally to all nations, regardless of their economic status, whether marked by impoverishment, affluence, or middle income. The 17 SDGs recognise the interconnectedness of different areas, acknowledging that actions in one sphere will have repercussions in others. This underscores the importance of development strategies that simultaneously address social, economic, and environmental sustainability. The SDGs embody an understanding that SD must encompass economic growth, social well-being, and environmental protection, with the overarching goal of leaving no one behind. The 17 SDGs constitute a visionary framework that unites nations under a common mission to uplift the most vulnerable communities. By addressing prevalent challenges such as gender discrimination, poverty, and educational inequality, these goals represent a commitment to building a better world for all. As a global blueprint for development, the SDGs emphasise a collective effort and a holistic approach that integrates the contributions of governments, businesses, non-profit organisations, individuals, and educational forums.

HEIs, in particular, assume a pivotal role in confronting and alleviating educational inequalities, fostering innovation, and advocating for a more equitable and sustainable world. Accomplishing these goals demands a blend of creativity, expertise, technology, and financial resources. Several studies emphasise the indispensable role of this sector in advancing SDGs (González-Torre & Suárez-Serrano, 2022; Leal Filho et al., 2021; Serafini et al., 2022). HEIs are crucial in achieving all the SDGs. The document titled "Getting started with the SDGs in Universities", created by the Sustainable Development Solutions Network (SDSN)—Australia/Pacific in partnership with the Australasian Campuses Towards Sustainability (ACTS) and the global SDSN Secretariat, underscores the significant responsibility universities bear in incorporating the SDGs into their teaching, research, operations, and leadership. The guide offers practical advice and illustrative examples to encourage additional efforts in this direction. It argues that "the SDGs will not be achieved without this sector" (Kestin et al., 2017, p. 8). The guide suggests the following steps for HEIs to initiate and strengthen their commitment to SDGs: (1) assess existing initiatives and practices; (2) develop internal expertise and foster a sense of responsibility for the SDGs; (3) identify key priorities, opportunities, and areas for improvement; (4) incorporate, execute, and ingrain the SDGs into university strategies, policies, and plans; (5) supervise, evaluate, and transparently communicate their initiatives in support of the SDGs. However, it acknowledges that there is no universally "correct" strategy for HEIs to engage with the SDGs. The approach a university adopts is contingent upon various factors, including its size, context, research or educational strengths, funding availability, values, priorities, and the unique needs of the communities it serves.

In accordance with the recommendations outlined in the guide, Galán-Muros (2023) has introduced comprehensive guidelines aimed at assisting HEIs in formalising their approach and enhancing their impact on SDGs. These guidelines encompass the following key steps: (0) conceptual alignment and rationale, (1) awareness raising, (2) leadership commitment, (3) stakeholders' engagement, (4) strategy design and drafting, (5) knowledge provision,

(6) mapping of current initiatives, (7) prioritisation of future initiatives, (8) provision of resources, (9) implementation, and (10) long-term sustainability. These principles suggest that adopting a systematic approach can accelerate the contributions of HEIs to the SDGs. It is imperative for these catalysts of change to assume a growing responsibility in assisting students to become responsible and engaged citizens. They need to impart a clear understanding of the impending sustainability challenges, playing a crucial role in equipping future professionals with the requisite skills and competencies to effectively address the complexities of a globalised and intricate landscape (García-Feijoo et al., 2020). HEIs globally are reconsidering their educational and research initiatives, as well as their broader societal contributions, aligning them with the SDGs (Fia et al., 2023; Jorgensen, 2019). The 17 SDGs, established by the UN, serve as a collective framework for addressing global challenges and fostering a more sustainable future. These goals encompass a comprehensive agenda, including (1) no poverty; (2) zero hunger; (3) good health and well-being; (4) quality education; (5) gender equality; (6) clean water and sanitation; (7) affordable and clean energy; (8) decent work and economic growth; (9) industry, innovation, and infrastructure; (10) reduced inequalities; (11) sustainable cities and communities; (12) responsible consumption and production; (13) climate action; (14) life below water; (15) life on land; (16) peace, justice and strong institutions; (17) partnerships. These goals encapsulate specific targets intended as a roadmap for guiding and measuring progress towards their attainment by 2030. It is imperative for every institution to incorporate the SDGs into their policies, practices, and decision-making processes for promoting the overarching global sustainability agenda and fostering positive impacts at both local and global scales.

Nevertheless, the SDGs pose challenges that are not easily navigable for any institution. HEIs may encounter difficulties in garnering the political determination and resources necessary for their achievement. There are also inquiries regarding the extent to which HEIs, fundamentally oriented towards the cultivation of knowledge and understanding, can and should be held responsible for such practical and immediate effects on the surrounding world (Chankseliani & McCowan, 2021). While these concerns indicate that incorporating SDGs into HEIs is a complex endeavour, HEIs can leverage their influence to advocate for policy shifts and create partnerships with governments and industries, thereby mobilising the necessary support for SDG implementation. Besides, HEIs can be viewed as an opportunity to align their core mission of knowledge cultivation with the pressing needs of the global community. By actively engaging in SD, HEIs can enhance their relevance and impact, demonstrating a commitment to addressing real-world issues while fulfilling their educational mandate. The challenges posed by the SDGs can be seen as a call to action for HEIs to embrace their role as agents of positive change and contribute meaningfully to a sustainable future. HEIs are not immune to the pressing concerns of our planet, and the implementation of the SDGs can empower them to have more significant contributions in a country.

The integration of the SDGs into HEIs has gained increased attention in recent years, as demonstrated by a growing body of research (Chankseliani & McCowan, 2021; Griebeler et al., 2022; Kioupi & Voulvoulis, 2020). Despite this growing interest, there remains a significant gap in both knowledge and empirical evidence in this field. Specifically, there is an urgent need to comprehensively document the various strategies and activities implemented by universities globally, especially within low- and middle-income countries (Chankseliani & McCowan, 2021; Fia et al., 2023). Furthermore, there is a scarcity of research examining HEIs' policies in relation to the SDGs. This gap highlights the importance of conducting studies that explore the extent to which reforms in HEIs align with the SDGs. This study aims to contribute to the existing literature by investigating the extent to which recent reforms in Algerian HEIs serve as indicators aligning with SDG4.

Education is indispensable for attaining the objectives delineated in the 2030 Agenda for SD. Among the 17 SDGs introduced by the UN in 2015, quality education is explicitly formulated as a standalone goal, denoted as SDG4. However, it is intricately interconnected with all other SDGs. SDG4 serves as a strategic framework designed to "ensure inclusive and equitable quality education and promote lifelong learning opportunities for all" (UNESCO, 2016). This goal signifies a global imperative aimed at instigating positive transformation, underscoring the crucial role of inclusive and equitable education in the pursuit of a sustainable world. It involves seven targets: 4.1) quality primary and secondary education for all, 4.2) early childhood and pre-primary education, 4.3) equal access to technical and vocational education and training (TVET) as well as HE, 4.4) skills for decent work, 4.5) gender equality and equal access for all, 4.6) youth and adult literacy, 4.7) SD and global citizenship (GC). It also includes three means of implementation: 4. A) safe and inclusive learning environments, 4.B) scholarships for HE, 4.C) professional development for teachers (UNESCO, 2018). Aside from 4.1, which addresses primary and secondary education, and 4.2, which focuses on pre-primary education, all other targets and methods of implementation are pertinent to HE. Target 4.3 is explicitly associated with HE.

For HEIs to effectively contribute to sustainability, as outlined in SDG4, it is crucial to ensure inclusive and fair access to affordable, high-quality education. This entails eliminating discriminatory practices and fostering knowledge and skills conducive to SD. Accomplishing this goal requires establishing inclusive and safe learning environments, along with enhancing support for scholarships and qualified teachers, especially in developing countries. HEIs are required to actively engage in shaping and leading the SDG4 agenda, instead of passively observing from the sidelines (Chankseliani & McCowan, 2021; Ferguson & Roofe, 2020). However, there is a paucity of research exploring how HEIs are engaging with SDG4. With this gap in mind, the aim of this book is to significantly contribute to studies on HEIs and SDGs. We intend to achieve this by investigating the Algerian context from the perspectives of teachers and students concerning the implementation of digitalisation and

EMI. Through this exploration, we aim to provide valuable insights and deepen the understanding of how SDG4 is pursued within Algerian HE, thereby enriching the academic discourse on the intersection of HE and SD.

2.4 Current Status of Sustainability in Algeria

It is undeniable that Algeria made some efforts to cope with its major issues of sustainability such as energy resources and their management. The 2023 report of the UN's SDG about Algeria shows improvement in attaining the 2030 goals, where there has been a significant increase in reducing poverty rates and increasing population access to renewable energy (Sustainable Development Report, 2023). The Algerian government has demonstrated a heightened commitment to SDGs, as evidenced by a surge in initiatives across the social, political, economic, and educational domains. An indicator of the government's commitment to sustainable resource management is its recent decision to join the BRICS organisation. This strategic move, announced earlier this year by President Abdelmadjid Tebboune during a press meeting in Algeria, emphasises the government's intent to leverage the benefits of BRICS collaboration for advancing the nation and addressing a multitude of challenges. President Tebboune explicitly stated that membership in BRICS is envisioned to propel the country forward and provide solutions to various issues. This decision not only reflects a proactive approach to global cooperation but also signifies a deliberate effort to align national development goals with broader international sustainability objectives. The government's active engagement with BRICS focuses its recognition of the interconnectedness between global collaboration and achieving SD milestones.

In contrast to the preceding administration, President Tebboune's cabinet has exhibited focused political plans to address local issues that directly impact the daily lives of citizens. According to Lagha and Bachi (2018), the pressing concern of vehicular emissions in Algeria, with the number of diesel-powered cars soaring to five million between 2001 and 2011, has contributed to increased pollution and associated health risks. The imperative for vehicles with lower emissions gained significant traction post-2019, prompting Algeria to impose import restrictions favouring environmentally friendly cars, particularly those from the European Union or China. A notable departure from the policies of President Bouteflika's tenure is the current administration's emphasis on constructing new smart cities as a crucial step towards sustainable energy (Boudiaf & Zeghichi, 2021). This strategic shift recognises a departure from the previous government's lack of prioritisation of such initiatives (Gherbi, 2012). President Tebboune's administration is actively steering the country towards a more environmentally conscious path, addressing both immediate concerns like vehicular emissions and laying the groundwork for sustainable urban development through the establishment of smart cities. Another illustration pertains to climate change and environmental conservation. In the past, the establishment of new cities overlooked environmental considerations, leading to issues like pollution, destruction of natural and rural landscapes, and unregulated disposal of waste (Gherbi, 2012). The government of President Tebboune launched recently new measures to protect the environment exemplified in waste disposal recycling according to the vision of 2035 and reducing climate change in Algeria through reducing emission (Ministère de l'Environnement et des Énergies Renouvelables, 2022a, b).

The realm of HE has experienced positive strides in its commitment to sustainability. Notably, several regulations were introduced to incorporate sustainable economic practices into educational programmes. By the close of 2022, ministerial decisions underscored the pivotal role of universities in driving economic progress and equipping the labour market with essential skills. One noteworthy development occurred in September 2020 when a new regulation was enacted, urging all Algerian universities to establish a national committee dedicated to overseeing "start-up" and "innovative project" initiatives, effectively functioning as business incubators within the university setting. The primary objective is to guide and support students harbouring innovative ideas, transforming them into viable economic projects and, potentially, small-scale enterprises. This initiative not only fosters an environment of entrepreneurship within educational institutions but also aligns academia with economic development, facilitating the transition of creative concepts into tangible contributions to the national economy.

The government's commitment to fostering sustainability is evident in various programmes it has implemented. Algeria's overarching policy framework is dedicated to continuously seeking sustainable solutions for the multitude of challenges it confronts across different domains. This commitment underscores the nation's proactive efforts to draw upon the experiences of leading countries in SD, with South Korea serving as a notable example. In 2012, Algeria actively engaged in the Knowledge Sharing Programme in Seoul, leveraging the Korean experience to enhance its national vision for 2030 (Korean Development Institute, 2013). Furthermore, Algeria's collaboration with China under the Second Five-Year Plan for Comprehensive Strategic Cooperation highlights the government's persistent endeavours to elevate sectors such as the economy, agriculture, technology, and health (Hadakha, 2022). These collaborative initiatives, alongside others, have played a pivotal role in shaping the government's policies, both domestically and internationally. They have significantly contributed to the formulation of strategies that define Algeria's position within the national and global frameworks, aligning with its steadfast commitment to sustainability.

2.4.1 Environment-Friendly Economic Policies

The initial attempt into SD economics by the government dates back to 1962, as exemplified by the Tripoli Programme. This early initiative saw the government embracing a communist plan to foster economics within the country (Hadjam & Tari, 2019). Subsequent developments in the form of the 1976 and 1986 charters, along with the policies implemented between 2001 and 2015, demonstrated a discernible trajectory of economic expansion. However, it is noteworthy that these endeavours (more particularly before the millennium) showed limited acknowledgement of their potential implications on the environment (Hadjam & Tari, 2019). Despite economic growth, the environmental considerations in these policies were not as prominently addressed, pointing towards an evolving landscape that requires a more comprehensive integration of sustainability principles into economic planning for a harmonious and enduring development trajectory. As a response to the UN' commitment to safeguarding the environment and fostering sustainable economic development, Algeria has recently implemented a series of legislative measures aimed at reducing economic impact on the environment. These laws are designed to encourage practices that align with SDGs, reflecting the nation's dedication to environmental preservation and responsible economic growth.

In 2002, Algeria established Le Conservatoire National des Formations à l'Environnement (CNFE), known as the National Conservatory of Environmental Training, with a primary educational objective to promote awareness about environmental issues. The CNFE's comprehensive training programme encompasses 17 major components. Mrs Bouali, the general director of the CNFE, emphasises that the institution operates under the supervision of the Ministry of Environment and Renewable Energies, aligning its efforts with the SD goals in Algeria. Similarly, Centre National des Technologies de Production plus Propres (CNTPP), or the National Centre for Cleaner Production Technologies, also falls under the purview of the Ministry of Environment and Renewable Energies. An integral focus of the CNTPP is facilitating the ecological transition towards a more sustainable industrial landscape in Algeria. The centre actively fosters collaborative research initiatives between companies and institutions, striving to establish environmentally friendly technologies. Furthermore, the CNTPP oversees international cooperation through national plans, such as the "42 actions pour développer des Modes de Consommation et de Production Durables au service" (42 actions to develop Sustainable Consumption and Production Patterns in service) spanning the years 2016 to 2030.

Furthermore, Centre de Développement des Énergies Renouvelables (Centre for Renewable Energy Development) is a publicly funded institution dedicated to scientific and technological pursuits. Its primary goal is to propel the scientific and technological progress of energy systems, particularly focusing on solar photovoltaic energy and environmental bioenergy. In addition to this centre, there are several other institutions committed to balancing economic growth with environmental considerations. Noteworthy among them are the National Climate Change Agency, National Coast Preserve, National Observatory for Environment and Sustainable Development, and National Centre for the Development of Biological Resources. It is notable that these institutions are not operating in isolation. Numerous cooperation agreements forged with public universities exemplify a collaborative approach towards establishing SD in Algeria. These partnerships facilitate ground-breaking research and innovation, laying the groundwork for a future where environmental consciousness and economic prosperity coexist harmoniously. Together, these entities form a dynamic network dedicated to creating a more sustainable and resilient future of the country.

2.4.2 *Socially Valued Citizens*

According to Leal Filho et al. (2022), social sustainability is the support of certain social values such as equity, inclusion, and justice to citizens. A socially sustainable society strives to ensure that all individuals have access to basic rights, opportunities, and resources, fostering a sense of belonging and active participation. This approach emphasises the importance of building strong social networks, valuing diversity, and creating a supportive environment that enables individuals to thrive collectively. Ultimately, social sustainability contributes to the creation of a resilient and harmonious community where everyone can lead fulfilling lives. We can assess the fulfilment of social sustainability by evaluating specific criteria and characteristics. According to Griessler and Littig (2005), among the indicators of establishing a good social sustainability are equal opportunities in education, income distribution, housing and health conditions, and gender equity. We would incorporate security and employment as additional indicators to successful social sustainability.

Algeria has demonstrated prompt responsiveness to address social sustainability requirements and enhance living standards by undertaking various initiatives in education, housing, and employment. In a significant stride towards educational enhancement, Algeria has recently unveiled its most substantial university teacher recruitment initiative in years, earmarking an impressive 7400 teaching positions for the upcoming academic year of 2023/2024 (Kodache, 2023). This initiative is poised to address a longstanding issue of staff shortages that had previously constrained the productivity of Algerian universities. With an existing faculty of over 75,000 (Kodache, 2023), the move signals a concerted effort to bridge the gaps in teaching and supervision that have impeded academic progress. Notably, the Higher Education Ministry has bolstered its budget for the year 2023, allocating over $23 million, a remarkable 16% increase compared to the previous year of 2022, according to the Higher Education Ministry website (MESRS, 2023a). This augmented financial commitment highlights Algeria's dedication to fortifying its HE sector, ensuring a more robust and well-resourced academic environment for the benefit of both students and faculty alike. In the realm of housing, the country embarked on an ambitious endeavour known as the one million house project spanning from 2005 to 2009 (Saturi, 2016). This initiative was meticulously designed to cater to the diverse needs of different social classes, delineating four distinct housing schemes. These include housing projects tailored for rural areas, those catering to low-income households, initiatives for intermediate-income groups, and specialised projects aimed at the high-income demographic. Through these targeted efforts, Algeria aims to not only meet housing needs but also to promote inclusive development and improve overall societal well-being.

Algeria has recently introduced a novel initiative aimed at alleviating the challenges associated with employability. Despite facing criticism for its perceived short-term impact, the unemployment grant programme is designed to provide comprehensive assistance to novice job seekers. Operated under the oversight of l'Agence Nationale de l'Emploi (National Employment Agency), this programme not only offers financial support but also serves as a platform to guide, inspire, and bolster individuals navigating their inaugural job search. At a distinct level, the realm of justice in Algeria has not yet attained the same prominence as SD is concerned. Human rights organisations, often viewed as potential threats by the government, face challenges in shedding light on issues, particularly those related to the crimes perpetrated during the black decade (Zeraoulia, 2020). This lack of visibility underscores a complex dynamic where the pursuit of justice and the recognition of past injustices are hindered by political considerations, posing a significant obstacle to the comprehensive development of the justice domain in the country.

2.4.3 *Adaptive Governing*

Once we question sustainable corporate governance (Cardoni & Kiseleva, 2023), it is important to note that political stability in Algeria helped in establishing good indicators of sustainability compared to pre-2019 era (see Werenfels, 2007). The stability in Algeria's political landscape has served as a foundation for fostering an environment conducive to sustainable business practices. The absence of frequent political upheavals and a relatively stable governance structure have provided corporations with a platform to focus on long-term sustainability goals (we can also cite the example of "le nouveau modèle Algérien de croissance économique à l'horizon 2030").[1] This stability is reflected in the ability of companies to formulate and implement sustainable strategies without the disruptions that political instability might bring. Moreover, political stability often translates into a clearer regulatory landscape. In Algeria, a stable political environment has likely facilitated the development and enforcement of regulations that encourage and, in some cases, mandate sustainable corporate practices, especially after 2019. This regulatory clarity not only guides businesses in their operations but also ensures that sustainable governance is integrated into the legal framework governing corporate behaviour. Furthermore, the alignment of political stability with sustainable corporate governance underscores the interconnectedness of political, economic, and environmental factors. In Algeria, this synergy has likely facilitated

[1] The model had been suggested by the Ministry of Finance, and it includes three phases: 2016–2019; 2020–2025; and 2026–2030. By 2030, the government should have facilitated more plans for a long-term SD and green economy.

collaborative efforts between the government and businesses to address sustainability challenges collectively. The establishment of partnerships and initiatives that promote environmental responsibility and social welfare may be more feasible in a politically stable environment.

The emergence of the new regime after 2019, in response to the Al Hirak movement, marked a significant turning point in Algeria's political landscape. Committed to justice and transparency, the new leadership ensured that the longstanding figures of the previous regime, which had governed the country for two decades, faced a fair trial. This unprecedented move aimed at holding them accountable for their misfeasance and addressing the grievances voiced by the Al Hirak movement. The trials became a focal point for the nation and helped to symbolise a break from the past and a commitment to upholding the principles of accountability and the rule of law towards a stable era in the Algerian troubled historical chapters. Although some (Davis, 2019; Kilavuz et al., 2023; Rachidi & Aballagh, 2023) may still think that Al Hirak prompted anti-democratic and silencing opposition methods, especially during the protests, it is important to acknowledge the political stability it brought for political sustainability which has other implications on social and environmental sustainability as discussed previously. In addition to the legal repercussions, the trials also served as a platform for a broader national dialogue on the need for political reform and systemic change. The public, having witnessed the accountability measures taken against the former leaders, became more actively engaged in shaping the future trajectory of the country. The outcomes of these trials not only addressed the immediate concerns raised by the Al Hirak movement but also are more likely to lay the groundwork for a more transparent and accountable governance structure. The new regime's commitment to justice and its willingness to confront the past misfeasance played a crucial role in rebuilding public trust and fostering a renewed sense of hope for a better and more sustainable Algeria.

Acknowledging Al Hirak's potential role in advancing sustainability visibility in Algeria post-2019, it is imperative to express sustained concern over concurrent practices that not only undermine democratic principles but also imperil the trajectory of sustainable practices within the realm of HE. As Sadiki and Saleh (2023) argue, the North African region grapples with the "degeneration of democratizations", a consequence of public movements failing to catalyse improvements in social conditions. This conundrum prompts a nuanced exploration into whether Algeria's sustainability efforts operate with a predilection towards economic considerations. Is the prioritisation of economic facets indicative of a two-tiered approach to sustainability within the Algerian context? Alternatively, one must scrutinise whether the sphere of HE in Algeria confronts a perceptual marginalisation, wherein its role in bolstering sustainable development is inadequately acknowledged, potentially hindering its efficacy as a cornerstone for a robust sustainable development agenda.

2.5 Challenges and Impairment of Sustainability in Higher Education

While the government has been actively working towards implementing SD in the economic and environmental sectors, there remains a critical need for increased attention and reforms in the HE sphere. Bouherar and Ghafsi (2021) highlights that despite ongoing efforts, the HE domain requires further enhancements to adequately address the evolving demands and challenges of the economy. It is important to acknowledge that some recent decisions appear to align with the economic objectives of the country; however, their effectiveness in achieving SD goals remains uncertain and warrants scrutiny. Thus, a comprehensive and strategic approach is imperative to bridge the existing gaps and ensure that educational reforms align seamlessly with the broader objectives of SD in both economic and environmental contexts.

We vehemently oppose the prevailing governmental strategy, which disproportionately sidelines the educational sector in favour of an exclusive focus on the economy. We advocate for a parallel bottom-up and top-down approach to facilitate comprehensive progress towards SD. The existing governmental approach predominantly adopts a top-down methodology, where the economy appears to exert its mechanisms across various sectors, with HE being no exception. This approach, however, has demonstrated its shortcomings in achieving enduring SD objectives. Its reliance on a singular perspective of sustainability, primarily centred on the economy, has proven ineffective, especially considering the recent shift from a heavily fossil fuel-dependent economic model (Aïssaoui, 2001). A bottom-up approach entails an educational system shaping the culture of the economy, thereby influencing how the economy is defined, a perspective that oversimplifies the reciprocal relationship between HE and the economy. Instances of a more balanced relationship between HE and the economy are illustrated as a parallel bottom-up and top-down approach. We can cite the example of Agasisti and Bertoletti (2022) which elaborates on their 18-year study examining 284 European regions, with HE as the primary variable. The findings they (2022) presented affirm that a region with a greater number of universities tends to experience more robust economic growth. According to Hanushek and Wößmann (2010), the significance of education in influencing economic prosperity has been a longstanding perspective. The quality of education a student receives in Ghana and Peru compared to Finland or Korea is not the same (Hanushek & Wößmann, 2010) arguing that human capital in the economy will also not be the same.

To implement a parallel bottom-up and top-down approach between HE and the economy in Algeria, addressing the educational challenges must take precedence. It is essential to resolve the educational turmoil before transitioning the sector to an economic model, especially in the absence of a solid foundational educational environment. One of the issues in Algerian HE is sustainability of the medium of instruction used in universities. According to Alkhateeb and Bouherar (2022), a significant number of graduates express dissatisfaction with their linguistic progress. This dissatisfaction carries broader implications for language, identity, and the sustainable linguistic experiences of these graduates. Linguistic unsustainability is attributed to language conflicts, as highlighted by Bouherar and Ghafsi (2021) in the context of language in education since independence. Striking a harmonious equilibrium becomes exceptionally challenging in HE due to the impairments of SD that we detail in the coming sections.

2.5.1 *Insufficient Institutional Support*

In 2023, the Minister of Higher Education and Scientific Research announced a 16% rise in the ministerial budget compared to the previous year, 2022 (MESRS, 2023a). Consequently, the total increase in the academic budget for the current year (2023) amounts to \$3.68 million, bringing the total budget to \$23 million. Compared to Saudi Arabia, for example, the national budget for education reached the limits of 50 billion dollars (Government Budget, 2023) as both the Ministry of Higher Education and the Ministry of Education have been merged together under one ministerial body (Aburizaizah, 2022). Recognising the inappropriateness of directly comparing income levels between Saudi Arabia and Algeria due to their differing economic statuses and SD goals, it is evident that they operate on distinct economic scales. According to the World Bank (2022), Algeria's Gross National Income (GNI) in 2022 amounted to \$187 billion, whereas Saudi Arabia's GNI stood at a significantly higher \$1.12 trillion. It is crucial to note that these figures underscore the substantial economic variations between the two nations. Examining the allocation of resources towards HE, however, Algeria earmarked a mere 0.01% of its expenditure to education in 2022. In contrast, Saudi Arabia demonstrated a more pronounced commitment by allocating approximately 4.46% of its total GNI to education. Notably, these percentages may vary slightly when compared to the officially "declared" national budgets of the respective governments. However, we opted to rely on the World Bank statistics for a more accurate representation, given the transparency inherent in these global figures as opposed to local reports on budgets or expenditures. This choice is driven by the enhanced reliability and objectivity associated with international sources like the World Bank. It is noteworthy that a nominal 0.01% allotment from the total Algerian GNI to the Ministry of Higher Education encompasses not only the requisite provisions for university maintenance but also the disbursement of salaries to the university staff. Upon closer examination of the remuneration component, it becomes evident that 62,286 (MESRS, 2022a) teachers and instructors contribute to the academic milieu of national universities in Algeria (Guessoum, 2019). Further extrapolating the financial implications, considering an average monthly remuneration of approximately \$500 per university staff member, the aggregate monthly salary outlay would exceed \$31 million.

Discussing the challenges posed by a limited budget in HE inevitably leads to questions about the infrastructure and resources of universities. Within the realm of implementing SD in Algerian HE, a significant hurdle lies in the struggle to effectively manage a high student population while maintaining the delivery of quality educational content. As highlighted by Guessoum (2019), the University of Science and Technology Houari Boumediene (USTHB) grapples with the pressure exerted by a large number of enrolled students on the availability of both infrastructures and teaching resources. This issue extends beyond USTHB and is reflective of the broader landscape of Algerian public universities, where a scarcity of decent infrastructures and teaching resources persists as a recurring challenge in most HEIs. The insufficiency of funding has multifaceted implications for the transformative aspirations of HE, often leading to trends that merely "enhance their current model, rather than change it" (Sebaaly, 2019, p. 176). Due to the inadequacy of financial resources, adaptation policies tend to align with existing financial frameworks. These adaptations involve addressing persistent issues with temporary solutions, particularly in response to the shortage of teaching equipment and facilities. For instance, Setif 2 University has implemented changes in the access of classrooms for the academic year 2023/2024, transitioning from a faculty-based staff approach to allowing all university staff involved in teaching. This strategic shift aims to mitigate the severe lack of infrastructures needed to accommodate the substantial influx of students. The overarching theme underlines the necessity for universities to navigate resource challenges and implement adaptive strategies within the constraints of limited financial support.

2.5.2 Short-Term Plans

Insufficient resources pose a significant barrier to attaining SD goals, yet this challenge represents just one facet of the complexities confronting Algerian HE. The inclination towards decisions aligned with short-term objectives is another contributing factor to the absence of a clear strategic plan for SD in Algeria. In 2019, the Minister of Higher Education initiated a referendum on adopting English as the medium of instruction, revealing overwhelming support, with more than 93% favouring the use of English over French in universities (Bouherar & Ghafsi, 2021). The period from 2019 to 2023 witnessed a heightened practicality in the decision to incorporate English, driven by ministerial imperatives, with plans to implement English as the primary medium of instruction for the academic year 2023/2024. The commitment to English language integration intensified further as the Ministry of National Education introduced English as a new subject in primary schools in 2022, signalling a deliberate shift away from the dominance of the French language. Amidst these developments, a pertinent question arises: why do we advocate for the implementation of English in both schools and universities? More than that, if we question the government' objectives on such a decision, then what purpose does English serve in class? The answer to this inquiry invariably revolves around the global prevalence and utility of the English language (Schneider, 2011), an influence that has asserted itself on Algeria as a nation (Jacob, 2019) that cannot afford isolation from the global community. Additionally, positioning English as a means to decolonise Algerian education from the linguistic heritage of French becomes evident (Bouherar & Ghafsi, 2021).

It is essential to acknowledge that decolonising education goes beyond merely substituting French with English; meaningful transformation necessitates a comprehensive overhaul that extends to the administrative framework currently dominated by Francophone personnel. Again, in Sebaaly's (2019) words, the emphasis should be on fundamentally altering the existing model rather than a superficial attempt at modification. The visibility of HE alone does not guarantee the success of such initiatives; instead, it has the potential to exacerbate challenges within the educational landscape. The complexity of the situation is further examined by the existence of double standards in the realities of HE. Notably, a substantial number of Algerian students, exceeding 30,000, whether privately or publicly funded, pursue education in France, constituting the second-largest foreign student diaspora in the French territories (Elbiled.net, 2018). Furthermore, despite the apparent shift in educational priorities, ongoing cooperation persists between Algeria and France. In 2022, both nations signed five cooperation agreements, with a focus on the HE sector (Algeria Press Service, 2022). Adding another layer of intricacy, the decision to introduce EMI in universities introduces a potential hurdle. This is especially true given that students accustomed to learning in French continue to grapple with fluency challenges in utilising the French lexical form (Bouherar & Ghafsi, 2021; Alkhateeb & Bouherar, 2022). The integration of English thus brings forth additional complexities, necessitating a nuanced approach to ensure effective implementation and address linguistic proficiency concerns.

An additional strategic move with a short-term progress towards SD goals involves the integration of digitalisation within the HE sector. Commencing the 2023/2024 academic year, Algeria marked a significant milestone by embracing a zero-paper policy in its HE sector, signalling a commitment to environmental sustainability. A pivotal element in this transition has been the ongoing creation and development of the PROGRES platform, designed to empower both educators and students with a versatile toolset. Notably, this platform serves various functions, including a grading system for students and monitoring the professional development of teachers like promotions and upgrades. The Minister of Higher Education's announcement on 22 September 2023, proclaiming the successful utilisation of digitalisation for student initial subscriptions (ENTV, 2023), resonates across multiple levels. To further envelop HE in a comprehensive digital framework, the ministry has introduced the Schéma Directeur Numérique (SDN). This strategic initiative comprises seven key axes, incorporating 16 programmes under these axes, and a total of 102 operational programmes scheduled for completion by December 2024 (MESRS, 2022b). Notably, the majority of these SDN programmes centre around streamlining administrative processes and enhancing the overall student experience within the university environment, far away from the realms of teaching and learning. This coordinated endeavour signifies a broader commitment to harnessing digital technologies for holistic advancement within the HE sector, but only if all aspects are taken into consideration.

The question of whether this initiative can be categorised as a short-term strategy poses a critical challenge to achieving long-term SD. Initially, the thrust of digitalisation in HE primarily targets administrative efficiency, with only a limited number of programmes directed at enhancing the actual teaching and learning processes. One of the examples to fortify this claim is the COVID-19 period. The global response to the COVID-19 pandemic exemplified a sudden and widespread shift from traditional face-to-face teaching to virtual platforms, compelling universities worldwide to hastily equip their stakeholders with requisite tools to navigate the challenges posed by quarantine measures. However, the Algerian experience deviates significantly from this global trend. The reliance on local e-learning platforms proved to be inadequate for accommodating the diverse needs of teachers striving to deliver effective online lessons. The platforms, designed primarily for uploading text-based or audio-visual materials, fell short of providing a comprehensive virtual learning experience. Moreover, the majority of educators resorted to free platforms such as Google Meet, as the university did not offer the paid Zoom version. Also, the pervasive issue of weak internet connectivity within university premises (Bellatrache & Aloutti, 2020), especially for teaching and learning purposes, further compounds the hurdles faced in aligning with SD objectives in the HE sector. Another formidable impediment to SD lies in the socio-economic disparities among students. The integration of technology into education often comes at a cost, and students from lower-income backgrounds may find it challenging to afford the necessary devices and internet connectivity, especially in rural areas. Remarkably, the university fails to address this digital divide by not providing computer rental services or dedicated spaces equipped with internet connectivity hubs for students. Consequently, individuals with limited financial means encounter substantial difficulties in accessing online learning resources. This predicament is substantiated by a study conducted by Lassoued et al., in (2020), which identified pedagogical, technical, and financial or organisational obstacles faced by students during the COVID-19-induced transition to distance learning in Algeria. The interplay of these challenges not only questions the viability of the current initiatives as short-term plans (i.e. EMI and digitalisation) but also emphasises the urgency of addressing these multifaceted issues to truly progress towards the implementation of SD in the HE landscape.

2.5.3 Less Investment in Human Development

Another enduring factor significantly influencing the trajectory of SD within Algerian HE lies in the realm of human capital. Although HE stands as a pivotal force in fostering human resources development, certain elements hinder its optimal contribution to the cultivation of human capital, particularly within the broader context of the Arab World governments (Karakhanyan, 2019).

The allocation of a $23 million budget for the coverage of HE expenses over the course of a year shows the negligence of the government towards this sector. This financial commitment unveils a complex scenario when scrutinising staff salaries and benefits within HEIs. In comparison to neighbouring nations like Morocco and Tunisia, the remuneration for HE teachers in Algeria is notably low. On average, a teacher receives around $500, encompassing all personal expenses, including housing. The inadequacy of financial incentives for educators is a predicament that advanced nations have continuously addressed ever since, recognising that securing the rights and well-being of educators directly correlates with the quality of teaching and, consequently, the overall standard of education (Dreer, 2022; Sutton & Wheatley, 2003). Contrasting this with the practices in the Middle East, we observe a comprehensive approach where, beyond financial compensations, universities ensure housing and annual holidays for their staff. This multifaceted support structure contributes to their rapid ascension in global university rankings, a phenomenon conspicuously absent from the discourse surrounding HE in Algeria. Addressing these issues is pivotal not only for the welfare of educators but also for the overall advancement of HE in Algeria. Initiatives to enhance financial incentives, secure housing, and acknowledge the critical role of educators could propel Algerian universities towards international prominence, fostering a conducive environment for SD initiatives. In this context, a nuanced consideration of human capital becomes imperative, reflecting not just an investment in education but a commitment to empowering individuals who will shape the future landscape of the nation.

The challenge of teacher well-being as an obstacle to educational improvement (Parker et al., 2012) parallels the potential detriment posed by a lack of comprehensive reforms to the overall quality of education. Within the realm of Algerian HE, an additional concern centres around the implementation of quality assurance measures. Guessoum (2019) elucidates that although the concept of quality assurance in Algerian HE was introduced in the 1980s, tangible strides were only taken in 2010 with the establishment of two committees: La Comité Nationale d'Evaluation (the National Evaluation Committee, CNE) and La Commission d'Implémentation d'un système d'Assurance Qualité dans les établissements d'Enseignement Supérieur (the Commission for the Implementation of a National Quality Assurance Framework in Higher Education Institutions, CIAQES). These committees lacked the necessary authority to enforce a serious commitment to quality assurance in HEIs. Consequently, internal reports submitted annually to the Ministry of Higher Education may lack credibility (Guessoum, 2019). Building on this, Hillman and Baydoun's (2019) research focuses on the critical relationship between auditing practices and the quality of academic assessments. Their (2019) findings suggest that reduced levels of academic auditing are associated with diminished quality assurance in academic contexts. Expanding the perspective to the broader context of corruption in the Arab World, especially in economically unstable countries like Algeria, universities also grapple with systemic challenges. Corruption infiltrates various aspects of academic institutions, including admissions, recruitments, evaluation systems, teachers' upgrades, and allocation of scholarships and resources (Hillman & Baydoun, 2019). This multifaceted challenge further emphasises the intricate interplay between institutional integrity, governance, and the pursuit of educational quality in the Algerian HE landscape and plays a significant role to slow the specious attempts for SD plans.

2.6 Sustainable Development Initiatives in Higher Education Across the Arab World

In this section, an examination of various SD models within the context of Arab HE is undertaken, with a particular emphasis on scrutinising Arab nations, including Qatar and the United Arab Emirates, as illustrative case studies. The Algerian government's adoption of a top-down strategy for the integration of SD in HE provokes interest, prompting a comparative analysis with other Gulf nations. A more nuanced exploration seeks to ascertain whether an alternative bottom-up approach is evident in these comparative contexts. While the Algerian government demonstrates a predilection for a top-down methodology, the effectiveness of such an approach warrants critical evaluation, especially in light of the observed practices in other Gulf countries. An essential dimension of this inquiry involves investigating the existence of a bottom-up dynamic, acknowledging the potential significance of grassroots involvement in driving SD initiatives within HEIs. An underlying premise of this discourse is the recognition that a top-down approach alone may prove insufficient in comprehensively addressing all facets of SD unless there exists a heightened awareness among faculty members and staff regarding their pivotal roles in advancing the objectives of SD within HE frameworks. Beyond the structural and administrative aspects, the success of SD implementation is contingent upon the active engagement and commitment of individuals at various levels within the academic community. Furthermore, an intriguing aspect of the Algerian HE case is the apparent utilisation of tokenism (Benouar, 2013; Guessoum, 2019) in pursuing international visibility for endeavours related to SD in HE. This policy of symbolic gestures raises questions about the depth of commitment and the genuine integration of SD principles into the educational fabric. A rigorous exploration of this phenomenon is essential to discern the extent to which tokenistic practices contribute substantially to the broader objectives of SD or merely serve as superficial endeavours for international recognition.

Occasionally, the impetus for cultivating sustainability awareness within society does not exclusively rely on the shoulders of academic faculty and university staff; it is also contingent upon the perspectives and policies articulated by the government. Fostering an understanding of sustainability in societal behavioural norms commences with the education of the younger generation (Alkhayyal et al., 2019). Thus, any SD initiatives envisaged by HEIs necessitate initiation at the early stages of students' education, with the university representing the culminating stage for the realisation of such objectives. Examining the case of EMI implementation in Algerian universities offers a pertinent illustration. Algeria embarked on the introduction of the English language in primary schools in 2022. However, the alignment of this linguistic behaviour among students with the incorporation of English at the university level has presented challenges. The introduction of English at the university level has induced confusion among both educators and students lacking proficient English language skills and therefore had to attend multiple trainings for that purpose. This unintended consequence may exacerbate difficulties rather than contribute to the resolution of challenges outlined in the SD plans emphasised by the Ministry of Higher Education. Hence, careful consideration and strategic planning are imperative to ensure that educational reforms align seamlessly with broader sustainability objectives at the level of universities. In essence, the impediments to SD in Algeria extend beyond the juxtaposition of national educational reforms and university-level initiatives, encompassing challenges intrinsic to the centralised governance structure. Bint El Hassan et al. (2021, p. 1) posit that "internally decentralizing higher education institutions … enable greater autonomy and academic freedom" within this context as a solution to long-term SD goals. An illuminating examination involves a comparative analysis of Algeria's approach to SD within HE against analogous cases in other Gulf countries, thereby providing a nuanced re-evaluation of Algeria's endeavours in this domain.

2.6.1 *Qatar SD Approach in Higher Education*

As per the Government Communications Office's (2023) delineation, Qatar aspires to metamorphose into a highly advanced society, fostering sustained progress and ensuring an elevated quality of life for its populace. The roadmap for this transformation is encapsulated in the National Vision of Qatar 2030, which articulates enduring objectives and provides a framework for the formulation of comprehensive strategies and execution plans. The National Vision employs a multifaceted approach, encompassing human development, social development, economic development, and environmental development as its four interconnected pillars (Government Communications Office, 2023). Central to the realisation of Qatar's National Vision is the pivotal role of education, serving as both a primary goal and a facilitator in the pursuit of SD (Sever & Tok, 2023). This reveals the integral connection between educational initiatives and the vision for the country. A deeper exploration of the thematic dimensions reveals that education in Qatar is intricately woven into the fabric of SD, with two themes serving as guiding principles according to Sever and Tok (2023). Firstly, education is positioned as a transformative force preparing Qatari citizens to transcend traditional economic paradigms, extending their perspectives beyond the confines of the nation's historical reliance on oil. In essence, it serves as a catalyst for fostering a broader economic and social consciousness among the populace, aligning with the diverse and dynamic global landscape. By instilling a multifaceted understanding of economic practices and social dynamics, the educational system becomes a cornerstone in shaping a citizenry capable of navigating the complexities of the contemporary world. Secondly, education is envisaged as a guardian of Qatar's national identity, religious affiliations, traditions, and cultural, ethical, and moral values. In this light, the educational framework assumes the responsibility of preserving and perpetuating the unique cultural heritage of Qatar. By nurturing a profound sense of national identity, education becomes a guardian of the nation's rich heritage, safeguarding it against the homogenising forces of globalisation. This dual function of education not only equips individuals with the skills needed for global engagement but also ensures the continuity of Qatar's distinctive cultural legacy. By emphasising economic diversification and the preservation of cultural identity, education emerges as a powerful instrument steering Qatar towards a future characterised by both prosperity and cultural resilience.

The thematic foundations discussed earlier manifest through a repertoire of instruments devised by the Qatari government to realise their educational SD goals. Among the pivotal instrumental mechanisms in ensuring the materialisation of SD in harmony with the National Vision is the Qatar Foundation (QF). Inaugurated in 1995 under the patronage of the Qatari Royal Family, QF serves as the epicentre for Education City—a confluence of education, research and innovation, community development, and SD initiatives. Within the educational sphere, Education City accommodates prestigious institutions such as Hamad Bin Khalifa University, Virginia Commonwealth University School of the Arts in Qatar, Weill Cornell Medicine-Qatar, Texas A&M University at Qatar, Carnegie Mellon University in Qatar, Georgetown University in Qatar, Northwestern University in Qatar, and HEC Paris. Beyond serving as a hub for educational institutions, QF assumes a multifaceted role as a strategic facilitator for research initiatives, contributing substantially to the realisation of both the National Vision and SD objectives. The foundation acts as a dynamic force in fostering research collaborations among institutions, creating an intellectual ecosystem that synergistically addresses the dynamic challenges in education and sustainable development. Education City, as facilitated by QF, represents a holistic approach to education, integrating diverse disciplines and global perspectives. This comprehensive educational environment is pivotal for nurturing a cohort of professionals and scholars equipped to confront intricate issues and advance the broader goals of SD. The amalgamation of various universities and institutions within Education City not only enhances the educational landscape but also fosters an atmosphere of innovation and interdisciplinary collaboration. Undoubtedly, QF is a result of ongoing efforts to levitate the standards and quality of education, and therefore Qatar came to a conclusion that "motivated and well-equipped teachers will provide high-quality education, and this will lead to accomplishing SD goals" (Al-Thani et al., 2021, p. 4). This conclusion was part of a long history of educational reforms. As part of Qatar's commitment to advancing quality education aligned with SD goals, persistent initiatives have been undertaken to reform educational paradigms for improved outcomes. The year 2002 witnessed the introduction of Education for New Era (EFNE), a pedagogical framework that centres around learner- and teacher-centric knowledge dynamics. Subsequently, the Supreme Educational Council and, ultimately, the Ministry of Education and Higher Education have played pivotal roles in operationalising a refined perspective on education (Al-Thani et al., 2021).

QF also sustains the government National Vision in promoting social and environmental SD through various programmes and innovations it hosts such as Qatar Environment and Energy Research Institute (QEERI) and Qatar Science and Technology Park (QSTP). The universities in the Education City deliver local environmentally friendly innovations that contribute to SD local goals (Al-Musalmani & Maalouf, 2023). Few examples to cite: the CIRS Department at Georgetown University focuses on environmental studies, exploring topics such as climate change, the political dynamics of natural resources, food security, water conflicts, and various other environmental issues; "Sahtak Awalan" initiated by Weill Cornell Medical College aims to discourage unhealthy habits and promote a healthy lifestyle and make informed decisions about food choices and lifestyle; Texas A&M University operates a sustainable water and energy initiative with the objective of strengthening collaborations between the energy and industrial sectors and research and educational institutions (Al-Musalmani & Maalouf, 2023). In addition to QF, the Education for a New Era (EFNE) initiative encompasses various dimensions, including educational services, curriculum development, teacher quality, and the accessibility of educational resources (Khodr, 2011), similar to CNE and CIAQES under the Ministry of Higher Education in Algeria. Furthermore, the UNESCO Associated Schools Network plays a contributory role in SD education in Qatar in integrating local and global sustainability issues into the curriculum through reinforcing interactive and participatory teaching and learning methodologies, fostering critical thinking, and eliciting changes in attitudes (Sever & Tok, 2023).

Amidst Qatar's concerted efforts to establish education for SD, there appears to be an increasing integration of SD practices within the educational community. As elucidated by Bruder et al. (2023), educators in both private and public schools in Qatar exhibit heightened involvement in education for SD. This signifies the adoption of a dual approach, encompassing both top-down and bottom-up methodologies in the context of SD in education. The top-down approach is highlighted by various governmental initiatives aimed at sustaining SD education, specifically in preparing the education sector to align with SD objectives in Qatar. This approach has garnered success among teachers and students, indicative of a bottom-up orientation. Noteworthy distinctions between the approaches adopted by Qatar and Algeria are different in terms of efficiency and inclusivity. Qatar's SD policies have demonstrated efficacy in meeting diverse local needs and are inclusive, targeting both policymakers and educators. In contrast, Algeria's approach diverges, manifesting efforts that lack efficiency and inclusivity, predominantly relying on a top-down approach in HE.

2.6.2 *Sustainability in UAE and Saudi Arabia Higher Education*

The role of HE in facilitating the integration of SD goal within the Middle East region extends across various spheres, encompassing universities, schools, and communities as exemplified through a diverse array of initiatives spanning university activities, faculty involvement, community partnerships, non-governmental organisations, research undertakings, and innovation hubs (Omar & El-Bastawissi, 2023). Moreover, within the Middle East, the United Arab Emirates (UAE) stands as a noteworthy participant in the pursuit of SD integration within its educational landscape. Notably, the establishment of HEIs in the UAE is relatively nascent (Jose & Chacko, 2017. This is particularly significant in light of the UAE's formation in 1971, signifying a context wherein the development of HEIs is linked to the nation's relatively recent history of Emirates reunion. The UAE's approach to HE is characterised by a federal system, overseen by regulatory bodies such as the Knowledge and Human Development Authority (KHDA) in Dubai and the Abu Dhabi Education Council (ADEC) (Jose & Chacko, 2017). These agencies serve as vital components of the educational infrastructure, contributing to the shaping and auditing of the UAE's educational landscape. Through these mechanisms, the UAE endeavours to align its educational system with global sustainability goals, emphasising a commitment to fostering holistic development within the UAE.

One of the embodiments of SD in UAE HE is Dubai International Academic City (DIAC). According to Abdullah (2020), within DIAC universities, approximately 20 pertinent sustainability-focused programmes are currently available. To mention a few examples, there is Middlesex University Dubai, which offers an MBA in Sustainability and Corporate Responsibility, along with an MA in Global Governance and Sustainable Development. The British University in Dubai extends its academic portfolio to include an EMBA in Sustainability, an MSc in Sustainable Design of the Built Environment, and a PhD in Architecture and Sustainable Environment. Additionally, Heriot-Watt University Dubai features an MSC programme in Renewable Energy, while Amity University Dubai presents a BTech programme in Solar & Alternate Energy. These diverse programmes reveal the commitment of DIAC universities to addressing sustainability challenges through specialised education serving UAE SD objectives. There is a high expectation for an increasing number of UAE universities to introduce sustainability-related programmes in the near future, fostering local sustainability efforts, bridging skills gaps, and promoting global engagement and information sharing (Abdullah, 2020).

Despite the adoption policy the UAE follows in education in general where policy borrowing seems to define the educational sector (Matsumoto, 2019;

O'Sullivan, 2013),[2] the country has been able to foster SD in national and international universities. The National Strategy for HE 2030 has been launched recently by the UAE government, and it contains four major pillars: quality, efficiency, innovation, and harmonisation through 33 key initiative plans (Gulf News, 2017). For quality, the government implements rigorous accreditation and encourages competition between institutions. For efficiency, the government motivates securing the productivity of HEIs. For innovation, the government also encourages an HE environment that produces creativity and innovation that would help the economy. Harmonisation has to do with creating a labour market competition among graduates. This last pillar, for instance, can be seen in the study conducted by Nuseir and El Refae (2021) where students from three universities—New York University, Abu Dhabi University, and Al Ain University—consider academic reputation, grants and funding, location and proximity, facilities and services, promotional and marketing channels as the main factors that influence their university choice.

In contrast to the UAE's federal system, which affords greater flexibility to its Emirates, the educational landscape in Saudi Arabia is heavily influenced by a centralised approach to education (Aburizaizah, 2022). This top-down approach finds its justification in the perceived threat from external institutions, a concern rooted in safeguarding the tribal social structure that characterises Saudi Arabian society. Additionally, the Saudi government appears to draw insights from neighbouring countries, particularly the UAE, where many concerns have been raised about the appropriateness of borrowing policy on the local society as we have discussed before. While some view centralisation as a constraint on SD plans, the government's attempts in HE demonstrate unique and promising efforts. Notably, the Saudi government allocated a substantial portion of its GDP to the educational sector in 2016, equivalent to one-third of the national GDP (Nurunnabi, 2017). However, Khayati and Selim (2019) argue that this financial commitment falls short of addressing the complex realities of HE in the country. Despite this, the financial investment reveals the Saudi government's dedication to advancing HE and investing in human capital, evident in its ambitious goal of expanding the number of universities in the kingdom by 230% since the turn of the millennium (Abubakar et al., 2020). This expansion initiative is complemented by the establishment of the National Commission for Academic Accreditation and Assessment (NCAAA) in 2004, later integrated into the National Center for Academic Accreditation and Evaluation in 2016, responsible for evaluating and accrediting local HEIs (Aburizaizah, 2022). The NCAAA serves as the government's oversight mechanism, ensuring the implementation of sustainable plans and

[2] We concur that the policy of borrowing in education and HE in particular has its consequences (see O'Sullivan, 2015; Tabari, 2014) and this is one of the drawbacks of SD model in UAE, but we believe that it is a leading model in securing SD in solving local social and economic challenges. One of the challenges faced by UAE is the Western threats of the imported educational policies to the Bedouin tradition of the country where UAE acted to maintain Abu Dhabi values, culture, and heritage under UAE vision of 2021 (Matsumoto, 2019).

programmes in HEIs and measuring their success in contributing to the economic and labour market. For instance, the King Fahd University of Petroleum and Minerals, with its industrial and economic ties to Aramco, plays a pivotal role in supplying graduates to the oil job market, while the Royal Commission for Jubail and Yanbu institution spearheads the development of petrochemicals and industrial cities. Despite the centralised nature of HE and the top-down approach to SD, the HE community, comprising both teachers and students, appears to embrace the government's sustainability goals. Alahmari et al. (2019) conducted a study revealing a positive sustainability awareness among faculty members. Another study by Abubakar et al. (2020) indicates that over 72% of Saudi universities are committed to implementing sustainability, with more than 53% establishing sustainability governance offices within their campuses. In aggregate, the sustained efforts of the Saudi government to foster practical platforms for SD in HE are beginning to yield positive responses from students and teachers alike.

2.7 Conclusion

This chapter has delved into the realm of sustainability to provide a comprehensive framework for analysing its principles within the context of contemporary reforms post-2019 in Algerian HE, focusing on digitalisation and the integration of EMI. Sustainability was initially introduced in the UN's seminal document "Our Common Future" in 1987, also known as the Brundtland Report, which asserts that global issues largely stem from unsustainable practices. Subsequent to its emergence, the concept of sustainability has been widely discussed across numerous disciplines, including educational studies. Sustainability, defined as the preservation of resources for future generations, remains a topic of debate, often encompassing three interconnected dimensions: economic, environmental, and societal. The integration of sustainability within the field of education has sparked diverse conceptualisations due to its debatable nature. This book predominantly employs the UN's concept of ESD, primarily because it was established as a universal imperative for all nations. Educational institutions, including HEIs, serve as crucial pillars in SD efforts, ensuring the fulfilment of present requirements while preserving the needs of future generations. These entities are instrumental in advancing the 17 SDGs outlined by the UN for global attainment by 2030. However, despite concerted efforts, the world continues to grapple with myriad challenges. The timeframe of the remaining six years appears insufficient for the thorough achievement of these goals, notably in securing universal human rights and ensuring the sustainable management of resources. Addressing these challenges requires collaborative efforts from various stakeholders, including researchers, to facilitate the transition towards sustainable living. This book seeks to contribute to this collective endeavour by focusing on recent reforms in Algerian HE, critically examining their alignment with SDG4.

SDG4 underscores the paramount importance of providing inclusive and equitable quality education, serving as a cornerstone for nurturing a sustainable global society. Algeria, particularly in the wake of Al Hirak movement and amidst the challenges posed by the COVID-19 pandemic, is steadfastly committed to advancing national development while concurrently pursuing global sustainability objectives through a multifaceted approach comprising various policies and programmes. These initiatives are aimed at fostering economic growth, enhancing social progress, and preserving the environment. Nevertheless, there exists a pressing need for intensified efforts to ensure a substantial impact on the SDGs. The pivotal role of HE in fostering sustainability is often overlooked, despite its well-documented potential as a catalyst for development. While HEIs are witnessing significant enrolment rates, there exists a pronounced deficiency in the support allocated to this sector. This inadequacy is especially evident in the limited availability of teaching and learning resources, which could potentially undermine the delivery of high-quality education. Moreover, the effort to decolonise education, marked by a shift from French to English as the primary medium of instruction while retaining French in administrative functions, creates a complex linguistic dynamic. The transition to English, despite its colonial heritage, and the implementation of digitalisation heralded as the "zero-paper policy", through platforms like PROGRES and Moodle, without adequate provision of necessary resources and literacy training for HEIs members, demands a thorough examination of progress towards SDG4. This goal emphasises the critical imperative of ensuring that education is accessible, equitable, and of high quality for everyone.

Despite Algeria's ongoing efforts to combat corruption through the digitalisation of HE, unfortunately, it still persists. Nepotism permeates various aspects, including university admissions, module and resource distribution, teacher timetables, transfers between universities, and student academic evaluations. These hurdles impede progress towards SDG4, which highlights justice and inclusiveness within the HE system. Furthermore, Algerian university teachers receive low incomes compared to many Arab countries, where housing, research, and holidays are often included in the salary package. Some teachers in Algeria endure long commutes on limited budgets to reach their workplaces, while others allocate a significant portion of their salaries to live near universities with their families. These challenges significantly impact academic performance and hinder the achievement of SDG4, which emphasises quality education. Prioritising the well-being of both staff and students is crucial for SD in Algeria's HE system, necessitating financial investments and the inclusion of educators' and students' perspectives.

Countries such as Qatar, the UAE, and Saudi Arabia have demonstrated significant progress towards achieving the SDGs by heavily investing in HE and fostering a strong commitment to sustainability within their academic communities. This highlights the importance for Algeria to adopt a more inclusive approach in its HE system, moving beyond its prevailing top-down model to integrate a bottom-up perspective, thereby increasing its contributions to the

SDGs. This chapter has established a holistic framework for examining the tenets of SDG4 within the Algerian context, with a particular focus on analysing recent reforms in HE practices and experiences.

References

Abubakar, I. R., Aina, Y. A., & Alshuwaikhat, H. M. (2020). Sustainable development at Saudi Arabian universities: An overview of institutional frameworks. *Sustainability, 12*(19), 1–15.

Aburizaizah, S. J. (2022). The role of quality assurance in Saudi higher education institutions. *International Journal of Educational Research Open, 3*, 1–8. https://doi.org/10.1016/j.ijedro.2022.100127

Abdullah, M. (2020). UAE education heads in the path of sustainability. *The Sustainabilist.* https://thesustainabilist.ae/uae-education-heads-in-the-path-of-sustainability/

Agasisti, T., & Bertoletti, A. (2022). Higher education and economic growth: A longitudinal study of European regions 2000–2017. *Socio Economic Planning Science, 81.* https://doi.org/10.1016/j.seps.2020.100940

Aïssaoui, A. M. (2001). *Algeria: The political economy of oil and gas.* Oxford University Press.

Alahmari, M., Issa, T., Issa, T., & Nau, S. Z. (2019). Faculty awareness of the economic and environmental benefits of augmented reality for sustainability in Saudi Arabian universities. *Journal of Cleaner Production, 226*, 259–269.

Aleixo, A. M., Leal, S., & Azeiteiro, U. M. (2016). Conceptualization of sustainable higher education institutions, roles, barriers, and challenges for sustainability: An exploratory study in Portugal. *Journal of Cleaner Production, 172*, 1664–1673.

Alfaki, I., & Ahmed, A. (2016). *From oil to knowledge: Transforming the United Arab Emirates into a knowledge-based economy.* Routledge.

Algeria Press Service. (2022). Presidents Tebboune, Macron chair signing ceremony of 5 bilateral cooperation agreements. *Online article.* https://www.aps.dz/en/algeria/44720-presidents-tebboune-macron-chair-signing-ceremony-of-5-bilateral-cooperation-agreements

Alismaiel, O. A. (2021). Using structural equation modeling to assess online learning systems' educational sustainability for university students. *Sustainability, 13*(24), 13565.

Alkhateeb, H., & Bouherar, S. (2022). Sustainability and educational language policy in Arab higher education: Findings from Q research. *Current Issues in Language Planning, 23*(5), 1–23.

Alkhayyal, B., Labib, W., Alsulaiman, T., & Abdelhadi, A. (2019). Analyzing sustainability awareness among higher education faculty members: A case study in Saudi Arabia. *Sustainability, 11*(23), 1–13.

Al-Musalmani, F., & Maalouf, S. (2023). Qatar Foundation: A sustainability innovator. In L. Cochrane & R. Al-Hababi (Eds.), *Sustainable Qatar: Social, political and environmental perspectives* (pp. 73–92). Springer.

Al-Thani, W. A., Ari, I., & Koç, M. (2021). Education as a critical factor of sustainability: Case study in Qatar from the teachers' development perspective. *Sustainability, 13*(20), 1–32.

Assalihee, M., & Boonsuk, Y. (2022). Reforming religion-based higher education for sustainable development: The case of Islamic studies international program at Prince of Songkla University. In M. Öztürk (Ed.), *Engagement with sustainable development in higher education: Universities as transformative spaces for sustainable futures* (pp. 67–82). Springer.

Ayres, R. U., Turton, H., & Casten, T. (2007). Energy efficiency, sustainability and economic growth. *Energy, 32*(5), 634–648.

Bader, E. E. (2005). Sustainable hotel business practices. *Journal of Retail & Leisure Property, 5*(1), 70–77.

Bandarage, A. (2013). *Sustainability and well-being: The middle path to environment, society and the economy.* Palgrave Macmillan.

Bansal, P., & DesJardine, M. R. (2014). Business sustainability: It is about time. *Strategic Organization, 12*(1), 70–78.

Bellatrache, R., & Aloutti, L. (2020). E-learning challenges in higher education institutions: Algeria case study. *Journal of Economic and Human Development, 11*(2), 264–276.

Benouar, D. (2013). Algerian experience in education, research and practice. *Procedia – Social and Behavioral Sciences, 102*, 361–367.

Bint El Hassan, S., Cordova, K. E., Rabadi, G., & Abu Elhaija, W. (2021). The (un) sustainability of higher education institutions in Jordan. *Frontiers in Sustainability, 2*, 1–6.

Birnbacher, D., & Thorseth, M. (2015). *The politics of sustainability: Philosophical perspectives.* Routledge.

Boeve-de Pauw, J., Gericke, N., Olsson, D., & Berglund, T. (2015). The effectiveness of education for sustainable development. *Sustainability, 7*(11), 15693–15717.

Boudiaf, S. M., & Zeghichi, S. (2021). New, smart, and sustainable cities in Algeria: Realities and challenges. Case of the city of Sidi Abdellah in Algiers region. In F. Trapani, N. Mohareb, F. Rosso, D. Kolokotsa, S. Maruthaveeran, & M. Ghoneem (Eds.), *Advanced studies in efficient environmental design and city planning* (pp. 209–224). Springer.

Bouherar, S., & Ghafsi, A. (2021). *Algerian languages in education: Conflicts and reconciliation.* Palgrave Macmillan.

Brinkmann, R. (2020). *Environmental sustainability in a time of change* (Vol. 279). Palgrave Macmillan.

Bruder, J., Hamwy, N., Sellami, A., & Romanowski, M. H. (2023). Examining Qatar's private and public-school teachers' readiness for teaching about sustainability. *Globalisation, Societies and Education, 21*(4). https://doi.org/10.1080/14767724.2023.2262404

Cardoni, A., & Kiseleva, E. (2023). *Sustainable governance: Concepts, metrics and contexts.* Springer.

Chankseliani, M., & McCowan, T. (2021). Higher education and the sustainable development goals. *Higher Education, 81*(1), 1–8.

Christensen, H. B., Hail, L., & Leuz, C. (2021). Mandatory CSR and sustainability reporting: Economic analysis and literature review. *Review of Accounting Studies, 26*(3), 1176–1248.

Cochrane, L., & Al-Hababi, R. (Eds.). (2023). *Sustainable Qatar: Social, political and environmental perspectives.* Springer.

Dale, A., & Newman, L. (2005). Sustainable development, education and literacy. International. *Journal of Sustainability in Higher Education, 6*(4), 351–362.

Davidson, M. (2010). Social sustainability and the city. *Geography Compass, 4*(7), 872–880.

Davis, M. H. (2019). The layers of history beneath Algeria's protests. *Current History, 118*(812), 337–342.

De Fine Licht, K., & Folland, A. (2019). Defining "social sustainability": Towards a sustainable solution to the conceptual confusion. *Etikk i Praksis-Nordic Journal of Applied Ethics, 2*, 21–39.

Dempsey, N., Bramley, G., Power, S., & Brown, C. (2011). The social dimension of sustainable development: Defining urban social sustainability. *Sustainable Development, 19*(5), 289–300.

Didham, R. (2018). *Education for sustainable development and the SDGs: Learning to act, learning to achieve. Policy brief; Advancing ESD policy, UNESCO global action program on education for sustainable development.* https://en.unesco.org/sites/default/files/gap_pn1_-_esd_and_the_sdgs_policy_brief_4_page_version.pdf

Disterheft, A., da Silva Caeiro, S. S. F., Ramos, M. R., & de Miranda Azeiteiro, U. M. (2012). Environmental Management Systems (EMS) implementation processes and practices in European higher education institutions–Top-down versus participatory approaches. *Journal of Cleaner Production, 31*, 80–90.

Dreer, B. (2022). Teacher well-being: Investigating the contributions of school climate and job crafting. *Cogent Education, 9*(1), 1–13.

Ekins, P. (2011). Environmental sustainability: From environmental valuation to the sustainability gap. *Progress in Physical Geography, 35*(5), 629–651.

Elbiled.net. (2018). Hadha huwa adad altalaba aljazayiriiyn fi aljamieat alfaransia (This is the number of Algerian students in French universities). *Online article.* https://www.elbilad.net/derniere-info/70542

Elliott, S. R. (2005). Sustainability: An economic perspective. *Resources, Conservation and Recycling, 44*(3), 263–277.

ENTV. (2023). liqa' khas mae wazir altaelim alali aleilmii alsayid kamal bidari (A special meeting with the Minister of Higher Scientific Education, Mr. Kamal Badari). *YouTube Video.* https://www.youtube.com/watch?v=-vz432me-_o

Ferguson, T., & Roofe, C. G. (2020). SDG 4 in higher education: Challenges and opportunities. *International Journal of Sustainability in Higher Education, 21*(5), 959–975.

Fia, M., Ghasemzadeh, K., & Paletta, A. (2023). How higher education institutions walk their talk on the 2030 agenda: A systematic literature review. *Higher Education Policy, 36*(3), 599–632.

Fischer, D., Jenssen, S., & Tappeser, V. (2015). Getting an empirical hold of the sustainable university: A comparative analysis of evaluation frameworks across 12 contemporary sustainability assessment tools. *Assessment & Evaluation in Higher Education, 40*(6), 785–800.

Galán-Muros, V. (2023). *General guidelines for the implementation of sustainability in higher education institutions.* https://www.un.org/sites/un2.un.org/files/unai_and_unesco_iesalc_guidelines_for_sdgs_implementation_final_version.pdf

García-Feijoo, M., Eizaguirre, A., & Rica-Aspiunza, A. (2020). Systematic review of sustainable-development-goal deployment in business schools. *Sustainability, 12*(1), 1–19.

Gherbi, M. (2012). Problematic of environment protection in Algerian cities. *Energy Procedia, 18*, 265–275.

Gichuhi, L., Obiero, C., & Kangethe, M. (2022). Developing a monitoring and evaluation framework for education systems in view of sustainable development goals: A case from Kenya. In M. Öztürk (Ed.), *Educational response, inclusion and empowerment for SDGs in emerging economies* (pp. 39–50). Springer.

Giddings, B., Hopwood, B., & O'brien, G. (2002). Environment, economy and society: Fitting them together into sustainable development. *Sustainable Development, 10*(4), 187–196.

González-Torre, P. L., & Suárez-Serrano, E. (2022). A framework for implementing and reporting United Nations sustainable development goals in Spanish higher education institutions. *International Journal of Sustainability in Higher Education, 23*(6), 1349–1365.

Goodland, R. (1995). The concept of environmental sustainability. *Annual Review of Ecology and Systematics, 26*(1), 1–24.

Government Budget. (2023). *Sectoral performance.* https://www.my.gov.sa/wps/portal/snp/aboutksa/governmentBudget/!ut/p/z0/04_Sj9CPykssy0xPLMnMz0vMAfIjo8zifT2dPQ38TQz9DUzDjA0CHV39TEy8XIwMzA30g1Pz9AuyHRUBPEIqgQ!!/

Government Communications Office. (2023). *Qatar National Vision 2030.* https://www.gco.gov.qa/en/about-qatar/national-vision2030/

Griebeler, J. S., Brandli, L. L., Salvia, A. L., Leal Filho, W., & Reginatto, G. (2022). Sustainable development goals: A framework for deploying indicators for higher education institutions. *International Journal of Sustainability in Higher Education, 23*(4), 887–914.

Griessler, E., & Littig, B. (2005). Social sustainability: A catchword between political pragmatism and social theory. *International Journal for Sustainable Development, 8*(1/2), 65–79.

Guessoum, A. (2019). Introducing quality assurance in Algerian higher education: The case of the University of Science and Technology Houari Boumediene. In A. Badran, E. Baydoun, & J. R. Hillman (Eds.), *Major challenges facing higher education in the Arab World: Quality assurance and relevance* (pp. 335–352). Springer.

Gulf News. (2017). *UAE launches National Strategy for Higher Education 2030.* https://gulfnews.com/uae/education/uae-launches-national-strategy-for-higher-education-2030-1.2097191#

Hadakha, A. (2022). *khuta khumasiat bayn aljazayir wa alsin.. madha sayajni albaladan min altaeawun al'iistiratiji?* (A five-year plan between Algeria and China. What will the two countries gain from strategic cooperation?). https://www.aljazeera.net/ebusiness/2022/11/15/خطة-خماسية-بين-الجزائر-والصين-ماذا

Hadjam, L., & Tari, S. (2019). Sustainable development in Algeria: Analytical reading of the concept and constraints. *Revue Recherches et Etudes en Développement, 06*(02), 121–140.

Hanushek, E. A., & Wößmann, L. (2010). Education and economic growth. In P. Peterson, E. Baker, & B. McGaw (Eds.), *International encyclopedia of education* (3rd ed., pp. 245–252). Elsevier.

Heinberg, R., & Lerch, D. (2010). What is sustainability? *The post carbon reader, 11*, 19. https://mycourses.aalto.fi/pluginfile.php/1159956/mod_page/content/9/Heinberg_WhatIsSustainability.pdf

Hillman, J. R., & Baydoun, E. (2019). Quality assurance and relevance in academia: A review. In A. Badran, E. Baydoun, & J. R. Hillman (Eds.), *Major challenges facing higher education in the Arab World: Quality assurance and relevance* (pp. 13–68). Springer.

Ikerd, J. E. (2012). *The essentials of economic sustainability.* Kumarian Press.

Jacob, C. (2019). 'Back to the "futur"': Mobility and immobility through English in Algeria. *Language & Communication, 68*, 6–16.

Jeronen, E. (2022). Sustainable education. In S. Idowu, R. Schmidpeter, N. Capaldi, L. Zu, M. Del Baldo, & R. Abreu (Eds.), *Encyclopedia of sustainable management* (pp. 3488–3497). Springer.

Jorgensen, T. (2019). Universities move to achieve the SDGs–and approach the next hurdle. *Implementing the 2030 Agenda at Higher Education Institutions: Challenges and Responses*, 44–89. https://www.hes-so.ch/data/documents/GLOBAL-UNIVERSITY-NETWORK-FOR-INNOVATION-12023.pdf

Jose, S., & Chacko, J. (2017). Building a sustainable higher education sector in the UAE. *International Journal of Educational Management, 31*(6), 752–765.

Karaarslan-Semiz, G. (2022). Outdoor education for sustainability with systems thinking perspective. In G. Karaarslan-Semiz (Ed.), *Education for sustainable development in primary and secondary schools: Pedagogical and practical approaches for teachers* (pp. 39–53). Springer.

Karakhanyan, S. (2019). Quality assurance in the Arab region in the era of customization: Where do we stand in terms of relevance? In A. Badran, E. Baydoun, & J. R. Hillman (Eds.), *Major challenges facing higher education in the Arab World: Quality assurance and relevance* (pp. 211–224). Springer.

Karatzoglou, B. (2013). An in-depth literature review of the evolving roles and contributions of universities to education for sustainable development. *Journal of Cleaner Production, 49*, 44–53.

Kestin, T., van den Belt, M., Denby, L., Ross, K., Thwaites, J., & Hawkes, M. (2017). Getting started with the SDGs in universities: A guide for universities, higher education institutions, and the academic sector (pp. 1–52). https://ap-unsdsn.org/wp-content/uploads/University-SDG-Guide_web.pdf

Khayati, A., & Selim, M. (2019). The status of innovation in Saudi universities. *Cogent Education, 6*(1), 1–12.

Khodr, H. (2011). The dynamics of international education in Qatar: Exploring the policy drivers behind the development of education city. *Journal of Emerging Trends in Educational Research and Policy Studies, 2*(6), 514–525.

Kilavuz, M. T., Grewal, S., & Kubinec, R. (2023). Ghosts of the Black Decade: How legacies of violence shaped Algeria's Hirak protests. *Journal of Peace Research, 60*(1), 9–25.

Kioupi, V., & Voulvoulis, N. (2020). Sustainable development goals (SDGs): Assessing the contribution of higher education programmes. *Sustainability, 12*(17). https://doi.org/10.3390/su12176701

Kodache, W. (2023). *Aljamieat aljazayiria ealaa bued khatwa min al alamia* (The Algerian university is a step away from internationalism). https://elikhbaria.dz/الجامعة-الجزائرية-على-بعد-خطوة-من-العا/

Korean Development Institute. (2013). *Establishment of Algeria's national vision 2030.* Korean Development Institute.

Lagha, H., & Bachi, A. (2018). Sustainable development in Algeria. *Algerian Journal of Environmental Science and Technology, 4*(2), 742–749.

Larimian, T., & Sadeghi, A. (2021). Measuring urban social sustainability: Scale development and validation. *Environment and Planning B: Urban Analytics and City Science, 48*(4), 621–637.

Lassoued, Z., Alhendawi, M., & Bashitialshaaer, R. (2020). An exploratory study of the obstacles for achieving quality in distance learning during the COVID-19 pandemic. *Education Science, 10*(9), 1–13.

Leal Filho, W., Frankenberger, F., Salvia, A. L., Azeiteiro, U., Alves, F., Castro, P., Will, M., Platje, J., Lovren, V. O., Brandli, L., Price, E., Doni, F., Mifsud, M., & Avila, L. V. (2021). A framework for the implementation of the sustainable development goals in university programmes. *Journal of Cleaner Production, 299*. https://doi.org/10.1016/j.jclepro.2021.126915

Leal Filho, W., Levesque, V., Sivapalan, S., Lange Salvia, A., Fritzen, B., Deckert, R., Kozlova, V., LeVasseur, T. J., Emblen-Perry, K., Azeiteiro, U. M., Paço, A., Borsari, B., & Shiel, C. (2022). Social values and sustainable development: Community experiences. *Environmental Sciences Europe, 34*(67), 1–13.

Li, W., & Yi, P. (2020). Assessment of city sustainability – Coupling coordinated development among economy, society and environment. *Journal of Cleaner Production, 256*. https://doi.org/10.1016/j.jclepro.2020.120453

Lozano, R., Ceulemans, K., Alonso-Almeida, M., Huisingh, D., Lozano, F. J., Waas, T., Lambrechts, W., Lukman, R., & Hugé, J. (2015). A review of commitment and implementation of sustainable development in higher education: Results from a worldwide survey. *Journal of Cleaner Production, 108*, 1–18.

Lozano, R., & Watson, M. K. (2013). Chemistry education for sustainability: Assessing the chemistry curricula at Cardiff University. *Educación Química, 24*(2), 184–192.

Lüdeke-Freund, F., Carroux, S., Joyce, A., Massa, L., & Breuer, H. (2018). The sustainable business model pattern taxonomy – 45 patterns to support sustainability-oriented business model innovation. *Sustainable Production and Consumption, 15*, 145–162.

Matsumoto, A. (2019). Literature review on education reform in the UAE. *International Journal of Educational Reform, 28*(1), 4–23.

McKeown, R., Hopkins, C. A., Rizi, R., & Chrystalbridge, M. (2002). *Education for sustainable development toolkit*. University of Tennessee.

MESRS. (2022a). *Le réseau universitaire* (University network). https://www.mesrs.dz/index.php/fr/agregats-2/

MESRS. (2022b). *Portail des plateformes numériques* (Digital platforms gate). https://www.mesrs.dz/index.php/fr/plateformes-mesrs/

MESRS. (2023a). *M. Baddari présent les grands axes du budget du secteur pour l'année 2023* (Mr. Baddari outlines the primary focus areas of the sector's budget for the year 2023). https://www.mesrs.dz/index.php/fr/2022/11/m-baddari-present-les-grands-axes-du-budget-du-secteur-pour-lannee-2023/

MESRS. (2023b). *Majmueat min alkararat alwizaria* (A set of ministerial decisions). https://www.mesrs.dz/index.php/2023/01/arretes-ministeriels-ar/

Meuleman, L. (2013). *Transgovernance: Advancing sustainability governance*. Springer Nature.

Ministère de l'Environnement et des Energies Renouvelables. (2022a). *Changements climatiques* (Climatic changes). https://www.me.gov.dz/fr/changements-climatiques/

Ministère de l'Environnement et des Energies Renouvelables. (2022b). *Stratégie nationale de gestion intégrée et de valorisation des déchets à l'horizon 2035* (National strategy for integrated management and recovery of waste by 2035). https://www.me.gov.dz/fr/environnement-urbain/dechets/

Moore, J. (2005). Is higher education ready for transformative learning? A question explored in the study of sustainability. *Journal of Transformative Education, 3*(1), 76–91.

Nurunnabi, M. (2017). Transformation from an oil-based economy to a knowledge-based economy in Saudi Arabia: The direction of Saudi Vision 2030. *Journal of the Knowledge Economy, 8*(2), 536–564.

Nuseir, M. T., & El Refae, G. A. (2021). Factors influencing the choice of studying at UAE universities: An empirical research on the adoption of educational marketing strategies. *Journal of Marketing for Higher Education, 32*(2), 215–237.

O'Sullivan, A. (2013). Globalized education policy flow – Taking "the best"? *Proceedings of the Second International Conference on Emerging Research Paradigms in Business and Social Sciences (ERPBSS) held in Dubai from 26-28 November 2013.* https://www.academia.edu/5381630/Globalized_Education_Policy_Flow_-_-_Taking_the_best_

O'Sullivan, K. (2015). Education reform in the UAE – Bringing private schools into the fold. *Journal of Teaching and Education, 4*(1), 311–320.

Omar, O., & El-Bastawissi, I. (2023). Bringing the SDGs to life by 2030: Best practices in higher education within the Middle East region. In W. R. Filho, I. R. Abubakar, I. D. Silva, R. Pretorius, & K. Tarabieh (Eds.), *SDGs in Africa and the Middle East region* (pp. 1–26). Springer.

Parker, P. D., Martin, A. J., Colmar, S., & Liem, G. A. (2012). Teachers' workplace well being: Exploring a process model of goal orientation, coping behavior, engagement, and burnout. *Teaching and Teacher Education, 28*(4), 503–513.

Patashnik, E. (2003). After the public interest prevails: The political sustainability of policy reform. *Governance, 16*(2), 203–234.

Rachidi, I., & Aballagh, A. (2023). Algeria's Hirak: Fading prospects for democratisation? *The Journal of North African Studies, 28*(6), 1515–1537.

Ruiz-Mallén, I., & Heras, M. (2020). What sustainability? Higher education institutions' pathways to reach the agenda 2030 goals. *Sustainability, 12*(4), 1290.

Sadiki, L., & Saleh, L. (2023). Degeneration and the demos in North Africa: Towards a 'critical' study of democratisation? *The Journal of North African Studies, 28*(6), 1444–1472.

Saturi, A. (2016). Altanmia almustadama fi aljazayar: alwaqie waltahadiyat (Sustainable development in Algeria: Reality and challenges). *Majalat Albahith, 16*(16), 299–311.

Schileo, G., & Grancini, G. (2021). Lead or no lead? Availability, toxicity, sustainability and environmental impact of lead-free perovskite solar cells. *Journal of Materials Chemistry C, 9*(1), 67–76.

Schneider, E. W. (2011). *English around the world: An introduction.* Cambridge University Press.

Scoones, I. (2016). The politics of sustainability and development. *Annual Review of Environment and Resources, 41*, 293–319.

Sebaaly, M. (2019). Digital transformation and quality, efficiency, and flexibility in Arab universities. In A. Badran, E. Baydoun, & J. R. Hillman (Eds.), *Major challenges facing higher education in the Arab World: Quality assurance and relevance* (pp. 167–178). Springer.

Serafini, P. G., de Moura, J. M., de Almeida, M. R., & de Rezende, J. F. D. (2022). Sustainable development goals in higher education institutions: A systematic literature review. *Journal of Cleaner Production, 370.* https://doi.org/10.1016/j.jclepro.2022.133473

Sever, S. D., & Tok, M. E. (2023). Education for sustainable development in Qatar. In L. Cochrane & R. Al-Hababi (Eds.), *Sustainable Qatar: Social, political and environmental perspectives* (pp. 329–347). Springer.

Sharma, A., Iyer, G. R., Mehrotra, A., & Krishnan, R. (2008). Sustainability and business-to-business marketing: A framework and implications. *Industrial Marketing Management, 39*(2), 330–341.

Sterling, S. (2001). *Sustainable education: Re-visioning learning and change* (Schumacher Briefings No. 6). Green Books, Ltd.

Sustainable Development Report. (2023). *Implementing the SDG stimulus: Includes the SDG index and dashboards.* https://s3.amazonaws.com/sustainabledevelopment.report/2023/sustainable-development-report-2023.pdf

Sutton, R. E., & Wheatley, K. F. (2003). Teachers' emotions and teaching: A review of the literature and directions for future research. *Educational Psychology Review, 15*, 327–358.

Tabari, R. (2014). *Education reform in the UAE: An investigation of teachers' views of change and factors impeding reforms in Ras Al Khaimah schools.* Sheikh Saud bin Saqr Al Qasimi Foundation for Policy Research.

The Brundtland Report. (1987). Report of the World Commission on Environment and Development: Our Common Future. Retrieved from: https://sustainabledevelopment.un.org/content/documents/5987our-common-future.pdf

The World Bank. (2022). *GNI (current US$) – Algeria, Saudi Arabia.* https://data.worldbank.org/indicator/NY.GNP.MKTP.CD?locations=DZ-SA

UNESCO. (2016). *Education 2030: Incheon declaration and framework for action for the implementation of sustainable development goal 4. Korea: UNESCO.* https://unesdoc.unesco.org/ark:/48223/pf0000245656

UNESCO. (2018). *Paving the road to education: A target-by-target Analysis of SDG 4 for Asia and the Pacific.* https://unesdoc.unesco.org/ark:/48223/pf0000265912

UNESCO. (2023). *What you need to know about education for sustainable development.* https://www.unesco.org/en/education-sustainable-development/need-know

Wade, R. (2008). Education for sustainability: Challenges and opportunities. In J. Coriddi (Ed.), *Policy & practice: A development education review* (pp. 30–48). NI Ltd.

Walton, J. (2000). Should monitoring be compulsory within voluntary environmental agreements? *Sustainable Development, 8*(3), 146–154.

Webb, A., & Rodriguez, L. (2022). Sustainable development goals and universities: Higher education engagement through the global master's in development practice programme. In M. Öztürk (Ed.), *Engagement with sustainable development in higher education: Universities as transformative spaces for sustainable futures* (pp. 1–18). Springer.

Werenfels, I. (2007). *Managing instability in Algeria: Elites and political change since 1995.* Routledge.

Zeraoulia, F. (2020). The memory of the Civil War in Algeria: Lessons from the past with reference to the Algerian Hirak. *Contemporary Review of the Middle East, 7*(1), 25–53.

CHAPTER 3

A Review of Algerian Higher Education Reforms, 1962–2019

3.1 Introduction

This chapter looks into an examination of the diverse reforms implemented by the Ministry of Higher Education under the oversight of successive Algerian governments and their respective political ideologies. Following its declaration of independence in 1962, the Algerian university underwent a series of transformative reforms that significantly reflected both its scientific pursuits and its ideological orientations, shaped by the varying policy perspectives of different governments. These governmental interventions played a pivotal role in regulating the educational landscape, aligning it with the socio-economic objectives of each administration in power. An illustrative instance is the critical higher education (HE) reform of 1971. This reform was instigated in the aftermath of Algeria's nationalisation of its oil and gas resources, which were previously under French control. The nationalisation marked a critical turning point in the nation's history, prompting a major overhaul of HE. The primary objective was to disentangle the Algerian university from lingering French influences and assert its autonomy. The impetus behind this educational transformation can be traced back to the larger context of opposing the French industrial and educational policies that had previously permeated Algeria. Between 1962 and 1970, there was a perceptible shift in perspective, wherein the French presence in HE was initially viewed as an avenue for cooperation. However, as the nationalisation measures unfolded, a re-evaluation occurred, leading to a reconsideration of the collaborative stance. An additional factor contributing to this discordance may stem from the divergent trajectories of communism pursued by Algeria and France, particularly in light of the fluctuating popularity experienced by the French Communist Party. Despite the Algerian government's ostensible alignment with communist ideologies endorsed by President Boumediene, his perspectives were incongruent with those of the Algerian Communist Party, whose political orientation emanated from the French

S. Bouherar, S. Salem, *Sustainability in Algerian Higher Education*, https://doi.org/10.1007/978-3-031-81347-4_3

Communist Party. In an article featured in the French newspaper *Le Monde*, Herreman (1965) expounds upon President Boumediene's government viewpoint regarding the Algerian Communist Party, elucidating its dissenting stance on the Algerian revolution and examining the relative immunity enjoyed by the party during the Ben Bella administration. Therefore, the evolution of the Algerian university is intricately intertwined with the nation's broader political and economic trajectory. The 1971 HE reform exemplifies how the Algerian government strategically utilised educational restructuring to assert its independence and chart a distinct socio-economic course. This dynamic interplay between national policy, resource management, and educational reform underscores the complex and evolving nature of the Algerian university system in response to the challenges of its post-independence era.

In response to the sweeping HE reform of 1971, the nation exhibited a discernible eagerness to reclaim its linguistic and cultural heritage within the domain of tertiary education. This chapter discusses this reform which was characterised by the abrupt substitution of the Arabic language for several years of prevalent French utilisation in higher education institutions (HEIs). Some experts (such as Benrabah, 2013; Le Roux, 2017; Bouherar & Ghafsi, 2021) have characterised this transformative initiative as a failure. Their critiques centre on the perceived inadequacy stemming from the absence of a thorough and all-encompassing analysis of the linguistic complexities inherent in Algerian society and its educational system. This oversight, they argue, has impeded the reform's efficacy and engendered challenges in reconciling the linguistic shift with the diverse educational landscape of the country. Having said that, however, it is crucial to acknowledge that the reform had a positive impact on addressing local labour demands and played a pivotal role in elevating the prominence of the Algerian economy. In response to the evolving industrial landscape of the 1970s, there was a comprehensive reconsideration of the structure of HEIs to align with the burgeoning industrial needs. A notable example is the University of Science and Technology Houari Boumediene (USTHB), which emerged as a key player in providing good quality education and producing highly skilled graduates who actively contribute to various national and international projects (Guessoum, 2019). In addition, the 1971 reform not only reshaped the landscape of HE but also opened up new avenues for sustainable development (SD) by placing a heightened emphasis on economic and social measures. The strategic alignment with industrial requirements showcased a forward-looking approach that fostered the generation of well-equipped professionals, thus reinforcing the vital link between HE and the broader economic and social fabric of Algeria.

A compelling argument can be made that the 1971 reform during the Boumediene administration strategically leveraged Pan-Arabism in educational restructuring, aiming to sever ties with France. While this move resulted in a linguistic conundrum, it concurrently opened up new economic vistas for Algeria. However, the intended trajectory faced unexpected turns, notably with the sudden death of Boumediene in 1978. Chadli Bendjedid assumed the presidency, bringing forth a new administration with supplementary educational perspectives to Boumediene's reforms. The chapter also discusses the impact of this administration on HE reforms that appears to be questionable, particularly in light of the developments in the late 1980s. Despite substantial government investments in both economic and educational reforms, there was a surge in unemployment rates during this period (Abid et al., 2023). This paradoxical outcome raises intriguing questions about the efficacy and alignment of the implemented reforms with the evolving needs of the Algerian society at that time. During this period, the landscape was further shaped by emerging movements advocating for national recognition, significantly influencing HE reforms. The Berber and Islamic movements, in particular, played a pivotal role and had a profound impact on the socio-political fabric of Algeria. These movements were not only catalysts for unrest but also key contributors to the complex socio-political dynamics of the era. Arguably, the Islamic group, FIS, is often attributed to being a significant factor in the tragic civil war that unfolded and persisted until the beginning of 2000s, when President Bouteflika assumed office and signalled the onset of a new chapter in Algerian history marked by the Concord Law.

Ultimately, this chapter explores the significant 2004 reforms that reshaped the structure of HE training besides the subsequent impact of glotto-political dynamics within Algerian society, particularly post-2019, on HE. The 2004 reform implemented a novel HE system, adopting the LMD (Licence Master Doctorat) framework, which streamlined university training by reducing the undergraduate level to three years and adjusting the master's and doctorate degrees to two and three years, respectively. This structural shift heightened the prevalence of the French language, especially in scientific disciplines (Rezig, 2011), thereby diminishing the role of Arabic and engendering a persistent linguistic conundrum in HE—a legacy that endures to this day (see Alkhateeb & Bouherar, 2022). The chapter concludes by delving into the complexities of post-2019 reforms, where the government appears to align itself with the demands of protesters, who also advocate for linguistic and cultural values besides political demands (Bentahar, 2020). Notably, this alignment includes a proposed shift from French to English as the medium of instruction (Maraf & Osam, 2022), introducing a new layer of intricacies in the ongoing evolution of HE in Algeria.

3.2 French-Algerian Cooperation amid Independence, 1962–1970

According to Matasci et al. (2020), the historical trajectory of colonial education in Algeria can be delineated into three discernible phases. Initially, during the early 1900s, education was predominantly orchestrated by missionaries of evangelisation (Little, 2005). This era emphasised the primacy of evangelisation over education, prompting scholarly inquiries into whether education was essentially a tool serving broader social and religious objectives (Bude, 1983). In addition, the educational landscape of this period primarily catered to economic exigencies, stratifying society into distinct classes, with the working class acquiring rudimentary linguistic skills to fulfil basic economic tasks (Matasci et al., 2020). The subsequent phase, characterised by a semblance of equality between settlers and the native population, witnessed initiatives such as the Service des Centres Sociaux in Algeria advocating for coexistence between Algerians and the French (Durham, 2020). This period also saw a surge in the expansion of schools and the modernisation of African colonies, spurred by the outcomes of the Brazzaville Conference in 1944 (Gardinier, 1988). The concluding phase, commencing in the 1960s, marked a pivotal shift in French colonial educational policy towards a competitive cooperation paradigm. During this period, France hastily extended academic assistance, while alternative avenues for collaboration opened up, reflecting the geopolitical dynamics of the Cold War era, with African colonies navigating between the influences of the Soviet Union and the USA (Matasci et al., 2020). In Manière's (2010, p. 164) own words:

> L'indépendance politique ne signifiait donc pas la fin des liens entre l'ancienne métropole et la partie africaine de son Empire. Le processus d'africanisation de l'enseignement devait se faire dans ce contexte particulier, à cheval entre les nouvelles ambitions nationales des États africains et une forte tendance centripète les rattachant au système français.

However, it is imperative to recognise that these cooperative endeavours were intrinsically linked to colonial power ideologies that profoundly shaped the perceptions of the native population regarding their land, culture, language, and identity. This relationship, exemplified in the case of Algeria, underscores the enduring impact of colonial legacies on the socio-cultural fabric of post-colonial societies. The intricacies of educational cooperation and its entanglement with colonial ideologies reveal a complex interplay between historical forces and the evolving aspirations of newly independent nations such as Algeria.

Indeed, the collaborative endeavours between these two nations did not emerge spontaneously or without precedent. Undoubtedly, the burden of achieving independence bore a considerable weight, requiring protracted efforts beyond the immediate temporal confines of a single day, month, or year. As of 1954, a mere eight years preceding Algeria's declaration of independence, the prevalence of illiteracy was alarmingly pervasive, attaining staggering rates of 98% among females and 94% among males (Tillion, 1958). The eve of independence witnessed a reprehensible assault on Algiers' library, by the French, specifically the Secret Armed Organisation (OAS), who not only plundered valuable manuscripts to relocate them to France but also ruthlessly incinerated the remaining books and historical documents (Bouherar & Ghafsi, 2021). This act of terrorism laid bare France's intentions concerning Algeria's educational heritage post-independence. Evidently, Algeria found itself bereft of its educational infrastructure, linguistic identity, cultural essence, and even historical legacy. The abrupt exodus of educators, a staggering 25,000 out of 27,000, as highlighted by Benrabah (2013), further exacerbated Algeria's vulnerability, rendering the nation in dire need of assistance, even if such aid were to emanate from France itself. This exodus not only weakened the fabric of Algeria's educational system but also underscored the deliberate destabilisation perpetuated by France, leaving the nascent nation grappling with multifaceted challenges in the aftermath of its hard-fought independence.

3.2.1 A French Protectorate Under Evian Agreements

As previously outlined, the objective was unequivocal: to establish Algeria as a French protectorate through a systematic policy aimed at depleting the country of its educators, experts, and engineers. The strategy was straightforward—initiate a significant influx of educators into Algeria before the independence, only to promptly withdraw in 1962, thereby triggering a sudden shortage. The campaign to gallicise Algeria (Khandriche, 1996) ultimately proved successful, rendering Algeria reliant on the French training force. On the eve of independence, France departed, taking a thriving workforce that had significantly contributed to the France development in Algeria. However, French interests in controlling Algeria did not leave Algerian borders. Foreseeing potential challenges, Algerian negotiators, represented by FLN leaders, proactively engaged in discussions with France to pre-emptively address the impending economic and educational challenges. Consequently, a series of agreements were forged between Algeria and France, designed to bolster the nation's economy and educational infrastructure. As anticipated, Algeria entered into an initial agreement with France (Algeria: France, 1962), allowing the secondment return of French teachers to mitigate the severe shortage precipitated by their departure. For example, the Evian Accords serve as a testament, explicitly outlined Chap. 2, section B, wherein it is declared that French personnel, including teachers and technicians, would be made available to the Algerian government (Algeria: France, 1962). Simultaneously, the shortage was addressed by enrolling Algerian students in French universities and colleges (Khandriche, 1996). This collaborative effort sought to mend the educational deficit and lay the foundation for Algeria's post-independence development.

We would confidently claim that the state of chaos began to manifest itself slowly between 1962 and 1965, as delineated by Revere (1971), who expounded on the division among elites based on secularism, religion, power dynamics, and legitimacy. The foundation for educational restructuring was laid upon four pivotal pillars: democratisation, nationalisation, Arabisation, and modernisation (Revere, 1971). However, the Algerian government during this period struggled to actualise these pillars effectively. The FLN party ascended to power, prioritising revolutionary legitimacy and side-lining the democratic element in governance. The secular socialist regime predominantly favoured an authoritarian approach to reclaim the nation's linguistic and cultural identity through educational reforms. Concerning nationalisation, the earnest efforts to nationalise education were hindered significantly by the strong influence of French educators and experts, as emphasised by Ferhat Abbas in his disputes with the ruling FLN party, which failed to deliver socialist education to the populace (Revere, 1971). The intent of Arabisation can be traced back to independence, with substantial reforms materialising in the 1970s, as French remained the dominant language in administration and education. Modernisation initiatives encompassed the expansion of schools and universities, many of which were extensions of the colonial educational system. For example, the Industrial Institute of Algeria, established in 1925 and later renamed the National Polytechnic School in 1984, along with Algiers University (established in 1909), Higher National Agronomic School (established in 1905 as the National Institute of Agronomic), and the National School of Administration (founded in 1945). The extension of HEIs included the establishment of the University of Oran 1 in 1967, the University of Constantine 1 in 1968, and the Institute of Agricultural Technology of Mostaganem in 1969.

Most of the collaborative efforts in education between the two nations transpired within the framework of the Evian agreements. Laskaris (2016) reports that an extensive cohort of more than 30,000 educators actively participated in Algerian HE, especially within Algiers University—a former French-administered institution prior to Algeria's independence. This substantial influx of "coopérants" in Algerian HE engendered the infusion of French cultural and social nuances into the Algerian milieu. Noteworthy unions, including the Professional Association of Higher Education in Algeria (APES), the National Union of Higher Education (SNESUP), and the Professional Association of French Teachers in Algeria (APIFA), were established by coopérants (Laskaris, 2016) as implicit conduits for this cultural exchange. Despite the explicit directives from President Boumediene and Minister of Education Ahmed Taleb, urging these coopérants to abstain from involvement in internal political affairs, and notwithstanding the Algerian government's vigilant attempts at monitoring them (Laskaris, 2016), it was, perhaps unbeknownst to them, that France wielded a considerable influence on Algeria. This influence was wielded through the strategic deployment of numerous French coopérants, instrumental in shaping the contemporary educational system, values, and culture of Algeria. Kadri (2014a) contends that HEIs in Algeria underwent a profound, albeit often unconscious, transformation shaped by the values and standards prevalent in French universities. This influence persisted, at least until the latter part of the 1980s. The Algerian government's pursuit of "Algerianisation" in education (Naylor, 2000, p. 63) was somewhat paradoxical, given their simultaneous reliance on French experts. This paradox is particularly noteworthy in light of recent accusations against French colonialism, confirming its involvement in generating instability within the Algerian context for more than a century. Despite the stated objective of fostering a distinctly Algerian educational system, the government's reliance on expertise from a source implicated in internal disturbances adds a layer of complexity to the dynamics shaping Algeria's post-independence HE landscape. This duality raises questions about the government's balancing act between asserting national identity through "Algerianisation" and navigating the challenges posed by external influences with potential destabilising effects.

Mauviel (2014) asserts that educational unions established by French coopérants in Algeria aimed to counteract the ramifications of French policies geared towards perpetuating dependency on French imperialism in formerly colonised territories. Naturally, this assertion rings all too familiar to Algerians, echoing the sentiments of a bygone era when the paternalistic refrain of "we came to educate and civilise you" was first uttered a century ago. It seems history has an uncanny way of replaying tragic narratives, and here we are, grappling once more with the same worn-out and ironic proclamation. Kadri (2014b) further notes that French coopérants at Algiers University, for instance, embraced the ideals of the French socialist movement against imperialism, disseminating this ideology among their students. However, in the formation of these unions, coopérants exhibited discriminatory practices against local educators, striving to maintain ultimate authority within educational institutions (Weber, 2014). Their influence during a decade-long pivotal period played an indispensable role in shaping the Algerian university system—a legacy that continues to be a focal point of ongoing efforts for improvement. Their imprint extended beyond the academic realm, permeating societal structures, contributing to the prevalence of Francophones in positions of influence within the country.

3.3 The 1971 Reform: Regaining Trust and Legacies

If we mention SD, shortly here, we would say that the actions of the French coopérants in Algeria reveal a nuanced agenda that extends beyond the ostensible focus on SD principles in education. While SD is ostensibly promoting education for all and equality in society, the underlying objective appears to be the subtle transmission of French values to the Algerian populace. It becomes apparent that economic progress, often associated with SD initiatives, is not the primary impetus; rather, the emphasis lies on instilling a cultural imprint through the educational system. The period of the war of liberation emerged as a challenging phase for the French coopérants, marked by staunch resistance from liberation movements opposing the assimilation of French values. The rejection of this cultural assimilation process further complicates the implicit goal of the coopérants. However, with the dawn of independence, Algerian leaders harnessed the soft power of education to facilitate a smoother realisation of the coopérants' objectives. The subsequent collaborative educational efforts between France and Algeria serve as a post-independence avenue to achieve what was once hindered by the wartime resistance. Beyond immediate temporal considerations, the French coopérants' vision for SD extends to a grander scale of time. Their aim is to cultivate a generation that, even after the coopérants' departure, would continue to align with French interests within Algeria. The potential interruption of this vision by President Boumediene's nationalisation acts in 1971 (detailed in this section) introduces a layer of uncertainty and is a subject of debate. Nonetheless, the coopérants appear to have achieved a measure of success in fostering Algerian Francophiles who later ascend to influential positions, thereby ensuring a lasting impact on the nation's trajectory. Within this narrative, President Boumediene emerges as a paradoxical figure. While he positioned himself as the architect of substantial HE reforms aimed at purifying Algerian education from lingering French influences, his concurrent actions granting the French control over various Algerian domains for a decade created a complex and contradictory portrait. This paradox reveals the intricate interplay between Boumediene's reformative aspirations and the pragmatic concessions made in the broader geopolitical context, challenging conventional expectations and offering a multifaceted perspective on his role in shaping Algeria's post-independence HE landscape.

To support our hypothesis, it is important to note that the majority of educators were coopérants, accounting for more than 86% of the total in Algiers University, even after President Boumediene announced his intentions to reform the Algerian university in 1971 (Kadri, 2014b). However, by 1976, they had departed, particularly at Algiers University (Kadri, 2014b). By examining the events that coincided with their departure from Algeria, we can see that Algeria began nationalising its gas and oil resources in the southern region. This move naturally upset the French and prompted them to react swiftly, attempting to oust countries that nationalise their resources from the international market (Musso, 2020). The conflict between Algeria and France extended to HE as well. Additionally, the rapid growth of Arabisation, particularly between 1974 and 1977 (Kadri, 2014b), compelled the French coopérants to leave. They felt that their linguistic and cultural, and more importantly economic, interests were being threatened and replaced. It is worth noting that the departure of the coopérants was more politically motivated than educational and it was no less politically significant than the HE reform that led to significant changes in the curricula, structure, and language of Algerian universities.

3.3.1 Restructuring Higher Education

The transformative national decree of 16 April 1976 marked a pivotal moment in the history of Algerian education, ushering in the most significant reform since the country gained independence. This change was particularly evident in the expansion of pre-tertiary education, characterised by the construction of new schools and a surge in student enrolment. The profound impact of this reform resonated throughout the education system, extending its influence to HE. It follows a logical progression that the substantial increase in primary and secondary students would naturally translate into a notable upswing at the university level. As a result, the reforms undertaken in HE were multifaceted, with a primary focus on decolonising Algerian universities from their French heritage. The objectives of the reform encompassed a comprehensive restructuring, including the expansion of institutions and universities, nationalisation of staff and curricula, and the promotion of equal access to HE for all citizens (Schoelen, 2023). Notably, the reform aimed to eradicate the physical presence of French coopérants from Algerian universities, signalling a deliberate effort to disentangle from colonial influences. It is worth noting that, despite the departure of the coopérants in 1976, their cultural and linguistic legacy continued to cast a shadow over the university Boumediene created. The emphasis on physically detaching the university from its French heritage, inadvertently, overlooked the enduring cultural values that would later manifest as significant challenges. Boumediene's approach, while successfully addressing the immediate need to remove the physical vestiges of French presence, neglected the deeper cultural dimensions that would prove to be contentious. In hindsight, it becomes apparent that the initial efforts to clear the university from the French influence might have benefited from a more nuanced consideration of the cultural landscape. This oversight, though well-intentioned, laid the groundwork for subsequent challenges tied to the persistence of cultural and linguistic values that continued to shape the identity of the university under Boumediene era.

Following the departure of the coopérants from Algeria, a significant chapter unfolded as the Algerian government set forth its ambitions to restructure the university system, a response aimed at effectively addressing the economic and social challenges confronting the nation. The thrust of the nationalisation of Algerian universities shifted away from being mere recipients of knowledge to a deliberate effort of integrating this knowledge into practical objectives, initially geared towards bolstering the economy. The restructuring of previously French-administered universities in Algeria underwent a profound transformation, including the creation of new HEIs. This transformation was marked by a diversification of subjects of study, as elucidated by Faradji (2015), particularly highlighting the development of sociology in Algerian universities since 1971. Notably, this reform, while broadening the academic landscape, did not place a significant emphasis on research but rather prioritised the cultivation of a skilled labour force to address the vacuum left by the departure of the French coopérants. Simultaneously, the professionalisation of HE played a pivotal role, opening avenues for technical ministries to independently tailor executive training programmes according to their specific needs (Ghouati, 2022). The establishment of the Ministry of Higher Education and Scientific Research in 1971, concurrent with these reforms, exemplifies this commitment to aligning educational objectives with the evolving demands of various sectors and supervising academics' training abroad. An illustration of the restructuring of HEIs is the establishment of the University of Science and Technology Houari Boumediene (USTHB). Alongside this university, several institutions underwent notable transformations, including the Industrial Institute of

Algeria, later renamed the National Polytechnic School in 1984, Algiers University, the National Institute of Agronomic, restructured into the Higher National Agronomic School, and the National School of Administration, along with Ecole Nationale Polytechnique. In tandem with these structural changes, the duration of academic training experienced a significant shift, with a four-year undergraduate programme, a two-year postgraduate programme leading to a Magister diploma, and a five-year PhD programme (Kouachi et al., 2019). Generally speaking, the years between 1969 and 1973 emerged as the pinnacle of constructing technological institutions, where graduates were not only immersed in academic learning but also obligated to undergo practical trainings in various companies. Those four years (1969–1973) signalled the start of a hands-on experience for graduates who were under the joint supervision of company leaders and university trainers, underscoring a holistic approach that aimed to bridge the gap between theoretical knowledge and real-world application (Ghouati, 2022).

The reform actively favoured the incorporation of practical subjects and technical studies, exhibiting a noticeable bias against the social sciences, which received less attention in comparison to the meticulous care directed towards technical disciplines. This dichotomy also mirrors the societal division that unfolded in the 1970s and 1980s, particularly between the urban proletarian class, deemed the primary labour force for companies' mobilisation, and the rural proletarian class, overlooked due to agricultural norms that failed to align with government interests. Consequently, the Algerian society witnessed a significant rural-to-urban migration, contributing to a socio-political phenomenon that exacerbated the disparities between societal classes. A parallel can be drawn in the university setting, where discernible signs of division surfaced. Hard science subjects experienced considerable success under this reform, creating a stark contrast with the less-prominent position of other social sciences. This trend aligns with the hypothesis proposed by Dreher et al. (2009), suggesting that the political ideologies of leaders play a pivotal role in shaping educational reforms. Dreher's study, encompassing 72 countries, including Algeria during the Boumediene era, underscores the interconnectedness between political ideologies and the educational landscape. The divergence in emphasis between hard sciences and social sciences within Algerian universities during this period resonates with Dreher and his colleagues in their cross-national study.

3.3.2 Arabisation

As part of the broader strategy for the nationalisation of curricula, Boumediene's administration in Algeria established the Office des Publications Universitaires. This institution, overseen by the Ministry of Higher Education, was founded in 1973 with the explicit mission of furnishing university students with teaching materials and textbooks and ensuring responsibilities for printing and distribution. Operating not only from its headquarters in Algiers but also through branches in Constantine and Oran, this institution served as a pivotal tool in advancing and executing the Arabisation[1] of curricula in HE. The Boumediene administration, and subsequently Chadli Bendjedid's government, employed a range of mechanisms beyond the Office des Publications Universitaires to facilitate the implementation of Arabisation. Notable among these were the High Council of National Language (established in 1980) and the Centres for the Intensive Teaching of Languages (established in 1981), as outlined by Benrabah (2005). The commitment to linguistic transformation persisted, with the formation of the Algerian Association for Arabisation in 1989 (Kadi, 2004). The foundational steps towards Arabisation were initiated with the issuance of the first decree ordering the Arabisation of HE on 25 August 1971, followed by the creation of the Permanent Committee for Higher Education (Benrabah, 2005). Another entity dedicated to supporting Arabisation within HE, the Permanent Commission for Arabisation, was established in 1980, further emphasising the government's commitment to linguistic reforms (Benrabah, 2005). Amidst the establishment of these institutions and committees, the Berber community expressed dissatisfaction with the increasing prominence of Arabic. This discontent manifested in resistance to Arabisation within their region, culminating in a period of instability during the Berber Spring violent events (Bouherar & Ghafsi, 2021). Notably, Berbers from the Kabyle region escalated their opposition, organising a school boycott for the academic year 1994/1995 as a form of protest against the ongoing Arabisation movements (Silverstein, 2003). The convergence of these efforts and resistance reveals the intricate dynamics surrounding linguistic policies in Algerian HE, reflecting a strong presence of linguistic diversity.

Despite persistent endeavours to implement Arabisation in HE, the trajectory of this process diverged markedly from the success witnessed in pre-tertiary education, where Arabisation effectively supplanted French (Le Roux, 2017). A conceivable explanation for this deviation lies in the resistance exhibited by French technocrats entrenched within the university system, potentially motivated by a perceived threat to their positions. This resistance becomes overly apparent through the outright denial of Arabisation by Berberism, categorising it as a clear and immediate menace necessitating elimination. A second factor impeding the swift integration of Arabisation in HE pertains to concerns about the potential marginalisation and isolation of universities and their graduates from progressive global trends (Le Roux, 2017). This apprehension gains further credence in the context of scientific disciplines such as medicine and engineering, where resistance to Arabisation has been notably pronounced (Mostari, 2004). Faced with such challenges, the government initiated a renewed push

[1] Arabisation began at the onset of independence, primarily focusing on primary schools. Subsequently, a comprehensive Arabisation plan was introduced in the 1970s and persisted until the 1980s (Abdulrazak, 1982). This initiative was primarily driven by the fact that the initial cohort of students in the Algerian education system after independence were the pioneers. Roughly 10–14 years after gaining independence, this first generation would have progressed to the university level, prompting the full implementation of Arabisation in the 1970s.

for Arabisation across all governmental institutions. This initiative commenced with the establishment of the High Council for Arabic Language in September 1998, bolstered by the legislative backing of the 17 December 1996 law, which mandated the use of Arabic in all government institutions, prohibiting the use of any other language for official purposes (Grandguillaume, 1997). Despite these concerted efforts to infuse Arabisation into HE, the impact was not uniform across disciplines. In social sciences, Arabisation found integral acceptance, albeit with only partial success in law, journalism, and psychology (Mostari, 2004; Benrabah, 2013), with 20% of the total graduate students taught in Arabic (Ruedy, 2005). However, the dominance of French persisted conspicuously in scientific and technical subjects such as medicine, engineering, and architecture (Rezig, 2011). Mostari's perspective, while somewhat conservative towards Hisb França[2] (The French Party in Algeria), prompted a more in-depth investigation by Bouherar and Ghafsi (2021). Their research delved into the components of such a party, revealing the involvement of Berberists and Francophones in Algeria and uncovering instances of their misconduct within the educational sphere. This intricate interplay of linguistic dynamics, resistance, and global considerations revealed the multifaceted challenges inherent in the pursuit of Arabisation and produced different ideological orientations within HE.

The schism within the realm of HE has been starkly delineated by divergent linguistic aspirations and pragmatic pursuits. This academic dichotomy is characterised by proponents aligning with the Arabisation movement on one side, while the opposing group, predominantly Francophones and Berberists, vehemently opposes it. A vivid illustration of this ideological clash is embodied in the stance of Abdelaziz Belkhadem, who held the position of Algeria's prime minister from 2006 to 2008. Belkhadem ardently advocated for a total Arabisation of the university curricula, concurrently launching a scathing attack on Francophones. In his accusatory rhetoric, he alleged that they were traitors to God and to the nation (Benrabah, 2013, p. 160). This episode serves as a testament to the failure of Arabisation in HE, attributable to the myopic neglect of SD in education. The government's decision to engage in a prolonged partnership with French coopérants for over a decade appears to have been a misguided strategy. What outcome did the authorities anticipate? Could they realistically expect a generation moulded by a French-centric education system to seamlessly espouse French values and thinking, yet remain unwaveringly loyal to Algeria? Moreover, the impracticality of reinforcing a language with its associated ideological movements[3] on a generation subjected to more than a decade of French indoctrination in HE becomes apparent. The current state of affairs is further complicated by the absence of comprehensive, long-

[2] Denoting individuals who are either publicly or secretly involved in the sinister French project against Algeria.

[3] See Mostari (2004), Benrabah (2005), Benrabah (2013), and Tilmatine (2015) for attributions of Arabisation to Islamic ideology.

term planning with well-defined SD objectives. This lack of foresight has significantly contributed to the intricate scenario at hand. Presently, we find ourselves entangled in a landscape marked by pronounced divisions between Francophones and Arabophones. Simultaneously, the schism extends beyond linguistic boundaries to encompass disparities between religious and secular groups, exerting a profound influence on the HE milieu over an extended period. The need for a nuanced and inclusive approach to education planning and language policy is acutely evident as we navigate the multifaceted challenges embedded in this complex educational terrain inherited by colonialism and exacerbated by post-independence governments.

3.4 Socio-economic Status: Higher Education Horizons Between 1980 and 2000

Nevertheless, the 1971 reform proved SD to be mismanaged, giving rise to repercussions rather than resolutions throughout the 1980s. Consequently, two opposing factions have emerged, both striving to dismantle Arabisation initiatives. Concurrently, the Algerian government embarked on an extensive Arabisation plan without due consideration for the inclusion of relevant ethnic groups. For instance, the *pieds noirs*, influenced by French indoctrination, adamantly refused to embrace Arabic, dismissing it as the language of the indigenous people with the lowest echelons of the social hierarchy (Benrabah, 2013). Ironically, the Algerian government, in implementing Arabisation, replicated this hierarchical structure. Through the Arabisation movement, social classes, particularly the Berbers, felt their linguistic rights were disregarded under Boumediene's language policies. Some scholars, however, such as Bouherar and Ghafsi (2021), contend that Berberists are mere puppets manipulated by France, insinuating that any linguistic or cultural movement raised by such a group is, logically, overseen by the French. This dynamic unfolds against the backdrop of French vigilance towards linguistic diversity that could potentially threaten the unity of French society, as exemplified by laws such as the loi Toubon and the Bas-Lauriol law (Uysal et al., 2008; Saulière, 2014; Garcia, 2014). Simultaneously, France appears to advocate for language rights for Berbers and other minority ethnic groups, aligning with Berber language and culture activists whom France hosts and supports. While seemingly divergent, this discourse on linguistic diversity resonates with the broader issue of language in Algerian HE. The Algerian government, in its pursuit of reforms, overlooked the crucial elements of language and identity. A foundational prerequisite for any educational reform is a meticulous settlement of the components of identity. Our identity shapes our aspirations and influences our approach to educational goals. Neglecting this reality or imposing an alternate reality in parallel can precipitate more problems than solutions. Addressing these foundational aspects is paramount for securing the long-term goals of SD.

Boumediene's misstep endeavour to enforce Arabisation upon the Kabyle region not only introduced heightened instability into his policies but also generated far-reaching consequences for Algerian education. While Boumediene effectively stifled the Berberist sentiments during his rule, the repercussions of his policies became evident in the early 1980s after his passing. Such repercussions can be exemplified in an event featuring Mouloud Mammeri, a prominent Berberist writer, at Tizi Ouzou University where he was expected to deliver a lecture about the Berber cultural heritage. The authorities' swift ban on the event escalated into confrontations with security forces, ultimately triggering the Berber Spring—a pivotal movement in Algerian socio-dynamics marked by resistance and disobedience against the ruling regime (Silverstein, 2003; Benrabah, 2005). During this turbulent period, Berberists, according to Bouherar and Ghafsi (2021), experienced a surge in cultural and identity movements. Kabyle songs, disseminating anti-Arab and anti-Arabic sentiments while glorifying Berberism, gained prominence, reflecting the heightened tensions surrounding linguistic and cultural identity. Simultaneously, the 1980s were characterised by the economic crisis of 1986 (Bélaïd & Abderrahmani, 2013; D'amato & Panta, 2017), setting the stage for the tragic events of October 1988, where hundreds of protestors lost their lives. The aftermath inadvertently opened the door for Islamist groups who were in the process of formulating a political ideology to govern the nation. However, the initial joy accompanying the newfound freedom and political achievements of the Islamists was short-lived. The state declared a coup against all Islamists in power and their political party, the Islamic Salvation Front (FIS in French), plunging the country into a protracted and devastating civil war between the secular army leaders and Islamist guerrillas (Bouherar & Ghafsi, 2021). This tumultuous period not only underscored the fragility of Boumediene's policies but also laid bare the deep-seated socio-cultural tensions and identity struggles within Algerian society, which would continue to shape the trajectory of the nation's educational and political landscape.

3.4.1 Validating the 1971 Reform

President Chadli Bendjedid followed a trajectory parallel to the reforms initiated by Boumediene, ensuring the sustained implementation of Arabisation within HE. The reforms of 1984 were designed to recalibrate the objectives set forth by the Boumediene administration, primarily centred on advancing the Arabisation movement and prioritising scientific and technical disciplines in university education. This restructuring marked a departure from the trends observed in the early 1970s when enrolment in social sciences constituted a dominant 67%, contrasting with technological subjects at 28% and medical studies at a mere 5% (MESRS, 2021). Nevertheless, a discernible shift occurred in the late 1970s and 1980s, introducing a conspicuous imbalance that favoured students in scientific and technical disciplines, while concurrently endorsing more practical majors within universities (Kouachi et al., 2019; MESRS, 2021).

In 1982 and 1984, the Bendjedid government unveiled additional reforms concentrating on organising the allocation of students in universities. This strategic move was geared towards prioritising technical subjects over social sciences, aligning the educational milieu with the exigencies of the labour market. The deliberate distribution policy manifested a clear asymmetry between technical pedagogical positions within universities and humanities. This nuanced approach ensured the cultivation of a workforce attuned to the government's ideological perspective. The HE planning policy, in steering the distribution of students, engendered an asymmetry, with social sciences seemingly neglected. Concurrently, the educational system began to favour individuals from affluent, influential families in education (Zartman, 1975), mirroring the ascendance of more secular military leaders to positions of authority. This selective approach resulted in the emergence of discriminatory educational practices within institutions, typified by the establishment of the "Bendali School (a private institution) for the education of political military and educational elites' children" (Benrabah, 2005, p. 429).

The initiation of university reforms at Boumediene had initially held control and authority over the university, but with the advent of Bendjedid and the economic crisis of 1986, that trend began to erode. The repercussions of the crisis, marked by soaring unemployment rates and a staggering external debt exceeding $16 billion by 1980, cast a shadow over Algeria (Ruedy, 2005). The Parliament uncovered pervasive corruption and mismanagement within public companies, which were entangled in political patronage (Ruedy, 2005). The year 1989 witnessed the surge of worker strikes, with 3389 strikes affecting 357,652 workers, extending beyond companies and government institutions to engulf universities (Alexander, 2002). Students played a significant role in channelling discontent towards what would become the next revolution. Unemployment soared to alarming levels, reaching 22% among workers and a staggering 70% among the persistently unemployed (Layachi, 2001). The university struggled to absorb its graduates into a workforce paralysed by the emerging political militancy on the horizon of the 1990s. In the midst of this turmoil, the Islamists ascended to power, posing a threat to the linguistic landscape of education. French and Tamazight languages faced jeopardy as Islamists advocated for the use of Arabic, citing religious legitimacy (Bouherar & Ghafsi, 2021). The Francophone militant elites, with support from France, assumed control, aiming to dismantle the FIS party and expel the Islamist authority from power (Bouherar & Ghafsi, 2021). This political turbulence marked the onset of a dark civil war that shattered any hopes for a better Algeria, undermining all efforts for peace. The nation grappled with the confluence of economic instability, political unrest, and linguistic tensions, setting the stage for a protracted and devastating period of internal conflict.

The university, seemingly impervious to the prevailing socio-economic upheavals such as the rise of Berberism, economic crisis, and civil war, remained steadfast in its commitment to the process of Arabisation. An example of this would be the revocation of Act N° 91-05, which was cancelled a few months ago in 1996. This led to the establishment of new deadlines for its nationwide implementation, specifying administrative sectors and universities to implement Arabisation by 5 July 1998 and 5 July 2000, respectively (Benrabah, 2005). Further reinforcing this trajectory, the High Council for Arabic Language was established on 26 September 1998 to ensure a comprehensive Arabisation process (Benrabah, 2005). However, the university policy towards Arabisation does not mean that it was unaffected by the surrounding events. This period witnessed a significant toll on the Algerian university, resulting in the erosion of its educators and scientific credibility. The impact was stark, with Islamic guerrillas specifically targeting teachers in schools (Ghanem, 2022) and universities, viewing them as complicit with the regime, deemed apostate by the guerrillas. Intelligence agents, aligned with the regime, also infiltrated the university, abducting staff perceived to be affiliated with the FIS party or sympathetic to its ideology. The ensuing chaos forced many (Colonna, 2004), educators included, to seek refuge in Europe and the USA, leading to a considerable brain drain and depletion of the university's intellectual resources. This tumultuous scene persisted in HE until the early 2000s, when President Abdelaziz Bouteflika's election marked the end of the war and ushered in new realities for the university. The impact of these transformative events left an indelible mark on the trajectory of HE in Algeria, shaping its evolution in the aftermath of a tumultuous period.

3.5 The LMD System in Higher Education: The 2004 Reform

Bouteflika's assumption of the presidency was widely perceived as a providential intervention to rescue Algeria from the shadows of the violent dark decade. During his inaugural visit to Tizi Ouzou in 1999, a Kabyle province that had been a focal point of dissent against the regime in the 1980s and 1990s, Bouteflika commenced his address with a pivotal declaration, stating, "[I]f you accept Arabisation … I accept to speak Tamazight and together we will be Arab-Berbers". This articulate manoeuvre aimed to underscore the imperative of fostering harmonious coexistence devoid of linguistic fervour as the key to navigating forthcoming challenges. Further underscoring his strategic positioning, Bouteflika continued, remarking, "[B]ut we entered an era (i.e. Boumediene's) of *les chefs d'etat stagiaires* … an era of meaningless egomania". His discerning critique of Boumediene's Pan-Arabism and the Arabisation movement policies revealed a nuanced disapproval, particularly resonant with the Kabyle region, which had perceived suppression under Boumediene's rule. Employing a satirical Algerian proverb that posits, "[I]f you ask a donkey about his father, he shall reply my uncle is a horse", Bouteflika deftly encapsulated the perceived absurdity of the prevailing political climate mostly inherited by Boumediene policies. Moreover, Bouteflika lamented the perceived decline of the university, characterising it as bereft of significance and lacking international recognition for its diplomas, in stark contrast to the 1980s when Algerian university credentials enjoyed global acknowledgement. This critical observation served to emphasise the erosion of Algeria's once esteemed educational standing and to underscore the necessity for a rejuvenated commitment to academic excellence on the international stage.

Upon assuming power, Bouteflika implemented the Civil Concord Law (Martinez, 2004; Zeraoulia, 2021) as a strategic measure to curtail the prevailing bloodshed. This legislative intervention aimed not only to quell the ongoing civil strife but also to persuade guerrilla forces to lay down their arms, offering them immunity from prosecution for past actions. The successful implementation of this law marked a definitive conclusion to a decade-long civil war, thereby ensuring the nation's security. Addressing the burgeoning Berberism movement, Bouteflika, in response to the Arouch Movement in 2001, acknowledged the significance of linguistic diversity by officially recognising the Berber language as the second national language, amending Article 3 of the constitution (Benrabah, 2005). This strategic move reflected a nuanced understanding of cultural dynamics within the nation and contributed to a more inclusive national identity. In the realm of pre-tertiary education, he displayed a commitment to comprehensive reform. In the year 2000, a committee of 158 experts was convened under the directive of Bouteflika to meticulously reassess the existing educational programmes (Khedouci, 2015). This initiative underscored his dedication to fostering a robust and adaptive educational framework aligned with contemporary needs. While his tenure witnessed commendable advancements, particularly in civil stability and cultural recognition, the economic sphere posed a formidable challenge. From 1999, the inception of his presidency, to 2002, the country grappled with economic intricacies shaped by military interests and characterised by a reliance on hydrocarbon rents (Harbi, 2002; Martin, 2003). This economic backdrop had far-reaching implications for subsequent HE reforms. At the end of his initial term, he unveiled a monumental reform in HE—the adoption of the Licence Master Doctorat (LMD) system. This transformation mirrored systems prevalent in Western European universities (Miliani, 2012). The introduction of the LMD system was emblematic of Bouteflika's aspiration to align Algeria's HE with international standards, despite the economic challenges that had persisted during his early years in office and persisted later on.

3.5.1 Restructuring the University Priorities and Trainings

The inspiration of the government's transformative efforts in HE stemmed primarily from a strategic objective to support the labour market with a more skilled workforce while concurrently dismantling the centralised control of student allocation in universities based on economic needs (Arabi, 2018). The adoption of the LMD system marked a paradigm shift, endowing educators with considerable flexibility in designing lesson content tailored to student needs and their own pedagogical insights. This transition also resulted in a reduction in the time instructors spent on evaluations compared to the previous standards in the old system. The goal was clear—to expedite the entry of a greater number of graduates into the labour market within the shortest possible time spent in the university. Beyond this primary objective, the LMD system includes additional aims, such as linking the university with SD goals, the decentralisation of universities from political and ideological constraints, and the facilitation of student exchange programmes to confer international recognition upon Algerian diplomas (Benziane & Boutaba, 2015). Operating on a credit-based and student-centric model (Benchicou et al., 2010), the system mandates students to accumulate a specified number of credits to progress to the subsequent academic year. At the undergraduate level (Licence, three years period), students are required to accumulate 180 credits to be eligible for postgraduate studies (Master, for two years). For a Master's degree, 120 credits are necessary for programme completion and potential qualification for a PhD through a national contest. For postgraduate studies (PhD, a minimum of three years), students are directed towards research-intensive training. The credit system is organised such that each academic year comprises two semesters, each offering 30 credits (Daghbouche, 2011), with various units to be taught. The transition to student-centred learning prompted a reorganisation of lesson and evaluation content, introducing innovative concepts such as competencies. In this revised pedagogical landscape, the teacher, once the focal point of the class, assumed the role of a guide in the learning process, with students actively generating teaching content based on the teacher's guidance. Project-based learning emerged as a fundamental component, empowering students to adopt the role of researchers rather than passive knowledge recipients. Another distinctive feature of the LMD system is its emphasis on the promotion of e-learning, internet utilisation, and technology integration in HE teaching and learning (Benchicou et al., 2010; Idri, 2012). Within this reformation, the growth of universities and polytechnics led to an augmented student population, with statistics from 2007 indicating 220,000 students according to Daghbouche (2011). While this reform appeared to be achieving at least one of its objectives, augmenting the skilled workforce in the labour market, it was also anticipated to generate significant challenges, as detailed later. Another notable dimension of this reformation pertained to the seamless exchange of students and faculty teachers among local universities (Daghbouche, 2011) and abroad. However, this endeavour encountered its set of challenges as the new reform was gradually generalised across the country. The LMD reform underwent continuous evaluation and refinement, with supplementary regulations introduced annually from 2004 to 2012, signifying a commitment to enhancing the efficacy of the system (Benziane & Boutaba, 2015).

The multifaceted objectives of the LMD system are underpinned by a desire to align HE with the evolving needs of the workforce and the globalised landscape. However, the transition to this system is not without its challenges and criticisms. While the focus on expediting entry into the labour market was evident, concerns are raised regarding the potential compromise of the depth and quality of education (Mahdjoub & Miliani, 2017). The accelerated pace of the system and the emphasis on credit accumulation raise questions about the thoroughness of academic exploration and the development of critical thinking skills. Moreover, the shift towards student-centred learning, though laudable in promoting active engagement, raises challenges related to the varying capacities and motivations of students to take charge of their educational trajectories "because the reforms were merely structural, with little focus on pedagogical issues" (Miliani, 2012, p. 219). The promotion of e-learning, internet utilisation, and technology integration brings its own set of opportunities and challenges. On one hand, technology opens up new avenues for access to educational resources, collaboration, and innovative teaching methods. On the other hand, it exacerbates existing disparities in access to technology and digital literacy. The surge in student enrolment, while indicative of increased access to HE, also places strains on infrastructure, faculty resources (Bouherar & Ghafsi, 2021), and the quality of educational experiences. The decentralisation of universities from political and ideological constraints was a significant departure from traditional models but presents its own set of complexities as well. While it aims to foster academic autonomy, there are concerns about potential disparities in educational standards across institutions and regions. The push for international recognition of Algerian diplomas through student exchange programmes reflected a global orientation but also raised questions about the compatibility of the LMD system with international academic frameworks and standards, especially with low teaching quality and research standards. The increased number of universities and polytechnics, coupled with the surge in student enrolment, reflects the government's commitment to expanding access to HE. However, it also highlights the need for parallel investments in infrastructure, faculty development, and quality assurance mechanisms to ensure a meaningful Education for Sustainable Development (ESD) experience for the growing student population. Furthermore, excessive emphasis on knowledge feeds into education's role in power dynamics, particularly when state controls the exclusive access to education (Khelfaoui, 2003). As a result, the objective to decentralise the university from the political ideologies of the state has never been attained in this reform. The ongoing evaluations and introduction of additional rules throughout the iterative LMD reform demonstrated a dedication to adaptability and responsiveness. Nevertheless, this approach tended to focus predominantly on coordinating conditions related to obtaining Licence, Master, and PhD, neglecting substantial social and economic challenges prompted by the reform.

3.5.2 Economic Aspirations

A foundational tenet of the LMD reform lies in its commitment to provide learners with high-quality training that aligns with the exigencies of the contemporary job market in Algeria. An instrumental means of realising this commitment is through the provision of scholarships. The Ministry of Higher Education, recognising a pronounced dearth of proficient educators in HE, inaugurated a specialised scholarship programme for Algerian students targeting prominent international destinations such as France, China, and the UK. In March 2014, a delegation comprising Algerian university representatives and high-ranking officials from the Ministry of Higher Education undertook a visit to the UK to engage in deliberations with regard to the placement of students for the ensuing academic year 2014/2015 (British Council, 2014). Notably, the British Council in Algeria played a pivotal role in assisting the process, with the initial meeting of British experts with Algerian students that took place in May 2014. This visionary initiative aimed to sponsor up to 500 Algerian students to pursue academic endeavours in British universities over five years (British Council, 2014; Belmihoub, 2018), commencing in 2014. Crucially, this pioneering endeavour sought to alleviate the acute shortage of English educators within Algerian universities. As individuals who participated in this transformative sojourn between 2015 and 2021, we find it imperative to elucidate the dynamics that transpired upon returning to Algeria. Regrettably, a substantial number of our co-sojourned colleagues opted not to return, electing to either remain in the UK or seek alternative opportunities beyond the confines of Algeria. While such decisions might be construed by some as acts of disloyalty to one's homeland, we, as colleagues, empathise with these choices. The rationale, we think, behind the decision of some individuals to forgo repatriation stems from a confluence of challenges encountered during the entire training period. The arduous process of scholarship renewal imposed considerable burdens on each sojourner, and the annual convocation of students in Algiers imparted a palpable sense that those who had pursued training abroad were not greeted with enthusiasm upon their return, to say the least. Moreover, the Ministry's recurrent threats to curtail financial support intensified the already daunting pressures faced by students. Notably, the psychological toll on students during their training period was exacerbated by the relentless demands of research activities. Those who did return to Algeria after completing their training were confronted with stark realities within HEIs. These included pervasive issues of incompetence among teachers, particularly those occupying leadership positions within university administrations. An unsettling atmosphere characterised by discrimination and inequity among educators further underscored the formidable challenges facing returning scholars in their efforts to contribute meaningfully to the enhancement of the Algerian HE landscape.

The correlation between the presented anecdote and the LMD reform lies in the observation that overseas training and student exchanges are typically governed by non-scientific conditions. These conditions adversely affect the quality of training and contribute to the loss of proficient educators, driven by a government-induced brain drain policy. The training courses abroad for teachers are usually governed by the impairment of the scientific committee and as a result negatively affect the quality of the process (Benchikh, 2017). Eventually, there is a consensus among central departments and faculties to acknowledge the deficiency in pedagogical training strategies for teachers

(Benchikh, 2017). This deficiency underscores the impracticality of achieving SD goals, as discussed in the preceding chapter. The manipulation of power within the university system directly hampers the execution of the reform. This is evidenced by the substantial corruption rates highlighted in Transparency International's evaluations, ranking Algeria 105th out of 178 countries in 2010 and 112th out of 182 countries in 2011 (Benrabah, 2013). Moreover, the economic objectives of the reform appear incongruent with long-term SD objectives for two principal reasons. First, language education quality is questioned, as a significant number of graduates express dissatisfaction with their linguistic skills (Arabic, French, English) at the university level, necessitating additional courses for market competitive employment (Alkhateeb & Bouherar, 2022). Second, the bifurcation between Arabic and French as mediums of instruction results in the job market favouring graduates instructed in French, particularly in scientific and technical fields (Le Roux, 2017). Overall, the impact of the reform on the labour market is a crucial aspect of its evaluation. The increase in the number of graduates entering the workforce requires examination of the quality and relevance of the skills imparted. While the reform addresses immediate skill requirements, the high number of graduates presents an ongoing challenge to the government's socio-economic reforms, as unemployment rates among university graduates remain high (Daghbouche, 2011). Additionally, employment opportunities often depend on nepotism and bribery rather than intellectual abilities (Mahdjoub & Miliani, 2017).

Largely, Daghbouche (2011) and Bouherar and Ghafsi (2021) cast doubt on the underlying motivations of educational reforms in Algeria, particularly regarding the country's adoption of the French HE system. The French HE, driven by the imperative to align university education with the demands of the French job market, faced challenges. In response to economic needs, the LMD reform in France sought to reduce the duration of university education. Paradoxically, Algeria, grappling with pre-existing unemployment issues, witnessed a worsening employment scenario due to the reform's impact that increased the number of graduates more than needed in the market (Bouherar & Ghafsi, 2021). The critical inquiry posed by Daghbouche (2011), "why do we have to stick to any French educational system?" (p. 468), prompts a nuanced exploration. This question transcends a simple answer and invites a deeper analysis. The essence of the reform is characterised by what can be termed as "myopic sustainability", where the primary objective seems to be the reform itself rather than addressing the actual challenges in HE through this reform. It appears as a means to quiet dissent within the population, a strategic move for electoral gains, and a tool to uphold political agendas. The reform falls short in terms of effectively addressing the intricate issues pertaining to HE and making meaningful strides towards achieving SD goals. As articulated by Mahdjoub and Miliani (2017), the "reality is somewhat distant from the rhetoric" of the regime. This indicates the need for a more comprehensive and purpose-driven approach in HE that transcends "merely structural reforms" (Miliani, 2012, p. 219), emphasising alignment with the substantial objectives outlined in the SD goals.

3.6 Higher Education After 2019

On 10 February 2019, the governmental pronouncement of paramount significance was made, thereby instigating a nationwide surge of discontent (Bentahar, 2020). The pivotal decision in question pertained to the pursuit of a fifth term in office by President Bouteflika, a move that carried the potential to extend his incumbency to an unprecedented 20 years. This development triggered an expeditious and forceful reaction from the public, manifested through widespread protests that collectively coalesced under the nomenclature "Al Hirak" (Bentahar, 2020; Kessar et al., 2021). The perplexity surrounding Bouteflika's bid for a fifth term was exacerbated by the incongruity between this decision and his earlier pronouncement in 2012 during a public address in Setif. In that instance, he invoked a Quranic verse proclaiming, "Allah does not burden a soul beyond that it can bear", thereby implying his perceived inability to navigate the challenges associated with a fourth term, let alone entertain the prospect of a fifth. The discordant power dynamics prevalent since 2009 were publicly acknowledged by the current Tebboune government, as elucidated by his Minister of Foreign Affairs, Ahmed Attaf, in a podcast interview with Aljazeera journalist Khadija Benguenna on 27 December 2023. Attaf asserted boldly that Algeria had endured a decade and a half devoid of a genuinely republican president (Atheer, 2023). Implicit within this proclamation was a critique of Bouteflika's protracted tenure, drawing parallels with historical instances where leaders disavowed the policies of their predecessors—a recurrent theme evidenced in Bouteflika' rejection of Boumediene's strategies and Boumediene's removal of Ben Bella. Under the burgeoning weight of public disapprobation, the government ultimately capitulated, officially announcing Bouteflika's renunciation of a fifth term. However, the persistence of protests compelled the establishment of formal elections in December 2019, culminating in the declaration of Abdelmadjid Tebboune as the nation's new president. This unrest permeated university campuses, where students participated in protracted demonstrations both during and subsequent to the presidential election, reflecting a profound dissatisfaction with the prevailing political landscape (Northey & Guemar, 2020).

The potential for transformative advancements in HE emerged with the prospect of significant objectives, yet the grievances voiced by protestors and students seemed inconsequential, failing to surpass the linguistic context of the university. In a study conducted by Maraf and Osam (2022) encompassing eight Algerian state universities, which included students actively engaged in Al Hirak movement for over a year, findings revealed a pressing demand for a transition from French to English as the medium of instruction in HE. Responding to this demand, the newly formed government under President Tebboune prepared to adopt English as a new language policy in the country. Upon assuming the role of Minister of Higher Education in 2019, Tayeb Bouzid initiated an online referendum regarding the incorporation of English in HE. The results indicated overwhelming support, with 93% of voters favouring the implementation of English in universities over French (Bouherar & Ghafsi, 2021). Another notable reform in HE post-2019 pertains to the digitalisation of administrative processes, with a little focus on e-learning. While the promotion of e-learning is not a novel concept, having been introduced in 1965 using television to teach students by the Ministry of Education (Louanchi, 1969), the 1990s and 2000s, renewed emphasis on these aspirations has surfaced since 2019. The ensuing paragraphs will provide a detailed examination of both initiatives.

3.6.1 English as a Medium of Instruction (EMI)

The inclination towards adopting English as a prominent language in Algeria began to manifest prominently in the aftermath of 2019. Notably, a pivotal shift in this trajectory occurred with the alteration of governmental institution names from French to English. This initial change set the stage for a comprehensive transformation in the educational landscape. The pivotal moment materialised in 2022 when President Tebboune, acknowledging the global significance of English, mandated its introduction as a subject in primary schools (Rouaba, 2022). It is noteworthy to draw parallels between this recent decision and a historical event from the 1990s. During that period, a similar choice was extended to parents, allowing them to opt between French and English as the medium of instruction in primary schools. This move, met with vehement opposition from the French lobby in Algeria, led to the dismissal of the Minister of National Education. An in-depth account of this incident was disclosed by Aljazeera, featuring an interview with Ali Ben Mohamed, who served as the Minister of National Education from 1990 to 1992. The widespread selection of English by parents prompted preparations for the recruitment of English teachers for the ensuing academic year; however, a deliberate leak of Baccalaureate exam questions compelled the minister to resign (Bouherar and Ghafsi, 2021). The ripple effect of incorporating English into the educational system did not cease at the primary level; rather, it permeated HE in the 2023/2024 academic year. This transition towards EMI at the university level was not arbitrary; it was influenced by both the demands of Al Hirak and the aspiration to equip students with linguistic skills conducive to global engagement. The government's commitment to this linguistic transformation is depicted by efforts to facilitate communication channels for scholarships in English-speaking countries. Entities such as the US embassy in Algeria have been instrumental in promoting scholarship programmes like the Study of the US Institutes for Global Student Leaders Programme and the Near East and South Asia Undergraduate Exchange Programme (Belmihoub, 2018). Also, the British Council's active involvement in endorsing English in Algerian HE is evident through a memorandum signed between the Ministry of Higher

Education and the British Embassy in 2021 (British Council, 2021). Moreover, the Algerian government has extended its encouragement beyond the academic realm, reaching out to various governmental institutions. Municipal employees, police personnel, civil protection agents, and other administrative workers have been afforded opportunities through agreements with Chevening scholarship issuers, facilitating leaves for academic pursuits in British universities.

In tandem with the shift towards English in HE, specific legislative measures have been enacted. Decree No. 1433, issued on 28 September 2022, mandated university chancellors to initiate preparations for the training of teachers in English, focusing on English for Specific Purposes (ESP) with a proficiency level of at least B2 or C1. Subsequently, a more detailed decree, No. 0005 dated 12 February 2023, urged chancellors to establish pedagogical committees to oversee the implementation progress of English in their respective universities. Additionally, on the same date, Decree No. 0006 was issued, outlining the formation of a national committee tasked with supervising the introduction of English across public universities in Algeria and overseeing the activities of local committees. The national committee comprises 18 members, led by the president. Furthermore, Decree No. 107, dated 22 March 2023, outlined the appointment of teachers responsible for curriculum design and IT engineers tasked with supporting the appointed educators throughout the implementation process.

In light of this linguistic paradigm shift, a remarkable observation surfaces. English, once viewed merely as an additional linguistic option, is now disrupting the longstanding belief that French exclusively serves as the gateway to the world, modernity, and advancements in science and technology (Benrabah, 2013). However, the successful realisation of such a linguistic transition necessitates substantial efforts. A comparative perspective can be drawn from a study by Rezig (2011), where attempts were made to correlate students' attitudes towards studying English with their final exam results. Astonishingly, the results revealed that 70% of students expressing positive attitudes towards studying English ended up failing their final exams. The attributed reasons primarily revolved around the lack of adequate teaching facilities and supervision. Drawing parallels to the decision to implement EMI in HE, the reluctance of the ministry to institute a comprehensive reform remains puzzling. While the incorporation of EMI may enhance international visibility for Algerian HE, concerns persist if graduates continue to lament the insufficient linguistic quality of Algerian universities (Alkhateeb and Bouherar, 2022). This dichotomy raises questions about the efficacy and sustainability of such linguistic transformations without a simultaneous focus on broader educational reforms to address the intricate linguistic challenges faced by both educators and students. Can the integration of EMI in HE yield enhanced research and teaching outcomes? In the absence of a comprehensive remuneration policy incentivising research achievements, the purpose of EMI remains nebulous, lacking a clear long-term strategic direction (SD policies). Consequently, the university finds itself ensnared by political ideologies, with direct repercussions on the educational landscape and a conspicuous absence of genuine intentions for reform. Boumediene's policy of Arabisation serves as a case in point, illuminating the political tensions with France, just as the post-2019 adoption of English signifies a crisis in Algerian-French relations post-Al Hirak. This prompts a critical inquiry into the enduring nature of these seemingly ostensibly inconsequential reforms and the perpetual assertions of unreal achievements in education on every election campaign. Is this a recurring trait intrinsic to every Algerian government?

3.6.2 *Digitalising Higher Education*

In order to ascertain the precise meanings of digitalisation and digitisation, it is important to establish clear definitions, a critical necessity for understanding the extent to which the government has implemented these processes. Digitalisation, as articulated by Gradillas and Thomas (2023), involves the remodelling of the socio-economic data through the adoption, application, and utilisation of digital artefacts. On the other hand, digitisation refers to designing digital artefacts through technical processes of conversion, representation, and enhancement (Gradillas & Thomas, 2023). HE embarked on a journey to digitalise its sector from early 2020, with a more detailed plan emerging in 2022. This strategic initiative includes 7 key axes, consisting of 16 programmes and a total of 102 operational programmes scheduled for completion by December 2024 (MESRS, 2022b). A significant development in this direction occurred on 23 October 2022 when decree N1401 was enacted by the Ministry of Higher Education. This decree established a national committee tasked with implementing and supervising e-learning in national universities. The primary objective is to promote e-learning and encourage the utilisation of university platforms for various purposes of distance learning, aligning with the broader aim of enhancing the quality of education and fostering a conducive environment for student learning (Abbas, 2023). Examining the Ministry's digital platforms, it is noteworthy that out of 41 platforms, only 1 is dedicated to students' academic pursuits—the e-learning platform (MESRS, 2022b), as highlighted in decree N1401. The remaining 40 platforms primarily serve administrative functions relevant to employees in administrative roles within the university. The core of online learning practices within universities involves pedagogically linked activities, lessons, and evaluations, with variations observed across institutions, particularly in the preference for scientific and technical subjects over theoretical subjects (Bessaid & Benyahia, 2023). Despite structural differences in websites, students' usage of e-learning platforms varies from one university to another (see, for instance Bouchrika et al., 2021 the case of University of Souk Ahras), a phenomenon we aim to explore under the theme discussed herein.

One prominent factor influencing this variability is the limited spread of internet connectivity within university campuses, where access is predominantly reserved for administrative purposes. Teachers and students lack readily available access points for the internet, except for desktop computers, which are sparsely distributed in specific areas of the university and insufficient to accommodate the large student population. Most staff rooms in public universities lack internet connectivity, extending to teaching classes and lecture amphitheatres. Moreover, there is a dearth of technology-integrated materials within teaching spaces and a lack of provisions for the maintenance of teaching materials in the event of technical issues. Additionally, the university does not facilitate the provision of personal computers through rental services or borrowing, nor does it offer well-equipped libraries with internet access and dedicated learning spaces, be they silent zones or common areas which are usually crowded and inappropriate. This absence of supportive infrastructure poses challenges to the digitalisation of HE spaces, necessitating alignment with economic and social changes. The persisting rise in unemployment and poverty rates since the pandemic (Ghanem, 2022) exacerbates the difficulty for students to acquire personal computers, especially when the university fails to provide such resources. The current educational landscape still involves teaching practices where it may be deemed inappropriate to ask students to bring their own computers into the classroom, given that a significant majority may lack the financial means to afford one.

The journey towards digitalisation in HE is an intricate process, entailing multifaceted challenges of the formulation and execution of strategic initiatives, as well as the provision of essential technological infrastructure. These disparities necessitate a comprehensive and inclusive approach to ensure the successful integration of digital technologies in the HE sector. Addressing these challenges demands specific commitment from educational institutions. However, it also calls for coordinated efforts from governmental bodies and policymakers. Bridging the existing gaps and creating an equitable and technologically advanced learning environment for all students requires a collective dedication to overcoming these obstacles. A cohesive strategy that combines the resources and expertise of various stakeholders is also essential to fostering a conducive environment for digitalisation in HE. This discussion not only marks the challenges of digitalisation but also illuminates the pivotal role that government perception plays in the implementation of SD in HE. It highlights the need for a harmonious alignment between government agendas, policies, and the evolving landscape of HE to ensure the seamless integration of SD principles. Despite the apparent resilience of universities in the face of socio-economic events in the 1990s, there is a discernible inability to link socio-economic plans with the outcomes of educational reforms. This recurrent pattern underscores the detrimental impact of myopic planning driven by short-term political considerations. Such planning fails to contribute meaningfully to the promotion of the long-term SD aspects in HE.

3.7 Conclusion

Since gaining independence, the promotion of HE for SD has posed a persistent challenge for successive Algerian governments. The fundamental tenets of SD have become deeply entangled with political agendas and conflicting ideologies, thereby influencing the manner in which education can effectively contribute to SD objectives. Economic liberalisation has exacerbated these concerns (Kabel, 2023), giving rise to tokenistic planning. The autonomy of HEIs has been compromised by political and military influences, preventing them from functioning with scientific legitimacy, particularly during the tumultuous periods of the 1980s and 1990s.

Holst (2023) delves into the concept of Whole Institution Approaches (WIAs), asserting that educational entities, such as universities, play a crucial role in advancing SD through practical concepts such as long-term commitment, human capacity building, and research. The commitment of universities to long-term goals has been hindered by their subjugation to the central government's vision, dictated by ideological orientations. Human capacity building within Algerian universities has witnessed significant deterioration. While offering sustainable professional development for university staff is a pivotal aspect conducive to education for SD (Rieckmann, 2018), Algerian staff trainings are influenced by unscientific factors, exemplified by the case of Algerian laureates in the UK. Moreover, educators within Algerian universities endure one of the lowest remunerations in the country, with a monthly salary of €230, often necessitating engagement in secondary employment (Bouherar and Ghafsi, 2021). Although the government, under President Tebboune's administration, has pledged a 47% salary increase for all governmental employees by 2024 (as Tebboune confirmed in his latest annual speech in front of parliament representatives on 25 December 2023), challenges persist. The inadequate recognition and encouragement of research within the Algerian university system remain significant impediments. For instance, the university fails to sponsor educators' participation in international conferences or financially support individual research endeavours. Ambiguities prevail concerning the privileges accorded to educators engaged in research-based activities versus those who are not. The restricted freedom of educators to explore sensitive topics, compounded by the influence of HE bodies, such as student unions, further exacerbates the situation. Overall, tokenism in SD within HE can be delineated through historical phases where both the government and the university system have faltered in adequately preparing HE to effectively contribute to long-term SD objectives.

Starting with post-independence events, Boumediene's government, in response to the abrupt departure of French educators, implemented a policy to enlist coopérants as a pragmatic solution. This strategic move successfully maintained commendable standards of teaching and ensured the production of high-quality graduates. However, this approach, although effective in some aspects, proved insufficient in safeguarding the national identity, which faced a crisis during the 1960s. Two primary factors emerged that undermined the alignment of HE with SD and exposed the transitory success of Boumediene's educational ideology. Firstly, the significant influence of coopérants played a pivotal role in shaping the landscape of HE. For an extended period, these foreign educators were instrumental in designing the trajectories of our future HE scenarios. While they undoubtedly contributed to the academic excellence of institutions, their role influenced the divergence from a more localised, context-specific approach to education and identity, thereby complicating the integration of Arabisation and ultimately SD principles. Secondly, the identity crisis prevailing during Boumediene's era further complicated the trajectory towards SD. The government's attempts to address the issue fell short, as Boumediene's policies not only failed to resolve the identity problem but, in fact, contributed to its exacerbation. The imposition of Arabisation without adequate language planning created challenges in fostering a cohesive national identity. This hasty and inadequately planned linguistic shift not only hindered effective communication within the education system but also perpetuated a sense of cultural dissonance, hindering the harmonious integration of diverse linguistic and cultural elements.

Despite the shortcomings of the Arabisation policy, it persisted as a defining feature of Algerian HE even after Boumediene's departure. The subsequent leaders, including Bendjedid, continued to align themselves with Boumediene's ideological objectives, with the university system during his era appearing largely unaffected by socio-economic dynamics. Instead, it continued the trajectory of reinforcing Arabisation without adequate language planning. This pattern persisted under President Zéroual as well, maintaining a continuity in policy without addressing the need for a comprehensive language strategy. In the prelude to the late 1990s, the government focus shifted towards attaining peace in the country and putting an end to the civil war. Unfortunately, little attention was directed towards achieving SD in HE or fostering an environment where education could actively contribute to broader SD objectives. The primary emphasis was on stabilising the nation, leaving the educational landscape relatively untouched in terms of substantial reforms aligned with SD principles. It was not until the early 2000s, particularly with the implementation of Bouteflika's Concord Law that brought about a semblance of peace, that a concerted effort was made to initiate substantial reforms in HE. At this juncture, the introduction of the LMD system marked a departure from the previous inertia. This shift aimed at modernising the HE structure, aligning it with international standards, and fostering a more dynamic and responsive learning environment.

The tokenism inherent in SD resurfaced again following the implementation of a HE reform, contributing to an influx of graduates into the job market beyond the norm. This influx exacerbated economic challenges and precipitated social instability. It was only under the leadership of President Tebboune in 2019 that a specific budget, allocating monthly remuneration, was earmarked for unemployed graduates. However, this measure, ostensibly aimed at resolving issues, ended up creating more problems. The myopic trajectory of the HE reform persisted with the transition from French to EMI, raising more uncertainties than providing clear solutions. This shift prompts educators to consider how they can linguistically prepare their students to navigate this challenge. Moreover, there arises a pressing need for special programmes to accommodate the substantial number of students, surpassing 1.6 million for the academic year 2022/2023 according to MESRS (2022a), affected by such a reform. These questions imply a lack of foresight and strategic planning, casting doubts on the sustainability motives and future prospects of HE in the delineation of coherent and sustainable horizons.

References

Abbas, A. (2023). The requirements of the directive plan for the digitization of the higher education sector: Reality and hope. *Journal of Legal and Economic Research, 6*(3), 271–283.

Abdulrazak, F. (1982). Arabization in Algeria. *MELA Notes, 26*, 22–43.

Abid, M., Benmeriem, M., Gheraia, Z., Sekrafi, H., Abdelli, H., & Meddah, A. (2023). Asymmetric effects of economy on unemployment in Algeria: Evidence from a non-linear ARDL approach. *Cogent Economics & Finance, 11*(1), 1–20.

Alexander, C. (2002). The architecture of militancy: Workers and the state in Algeria, 1970–1990. *Comparative Politics, 34*(3), 315–335.

Algeria: France-Algeria Independence Agreements (Evian Agreements). (1962). *International Legal Materials, 1*(2): 214–230. http://www.jstor.org/stable/20689578

Alkhateeb, H., & Bouherar, S. (2022). Sustainability and educational language policy in Arab higher education: Findings from Q research. *Current Issues in Language Planning, 23*(5), 1–23.

Arabi, M. (2018). Adoption of (LMD) system in Algerian higher education institutions: Response of international trends or a quality culture of higher education? *Innovation Creativity, 8*(1), 230–244.

Atheer. (2023). Qira'at fi almashhad aljazayri mae wazir alkharijia aljazayri Ahmad Attaf – budkast dhu shaan (A reading of the Algerian scene with the Algerian Foreign Minister Ahmed Attaf – dhu shaan podcast). *YouTube Video*. https://www.youtube.com/watch?v=Yd-i-EJYZBw&t=1164s

Bélaïd, F., & Abderrahmani, F. (2013). Electricity consumption and economic growth in Algeria: A multivariate causality analysis in the presence of structural change. *Energy Policy, 55*, 286–295.

Belmihoub, K. (2018). English in a multilingual Algeria. *World Englishes, 37*(2), 207–227.

Benchicou, S., Aichouni, M., & Nehari, D. (2010). E-learning in engineering education: A theoretical and empirical study of the Algerian higher education institution. *European Journal of Engineering Education, 35*(3), 325–343.

Benchikh, M. (2017). Governance of higher education institutions in Algeria: Inventory and assessment. In A. Farazmand (Ed.), *Global encyclopedia of public administration, public policy, and governance* (pp. 1–7). Springer.

Benrabah, M. (2005). Language planning situation in Algeria. *Current Issues in Language Planning, 6*(4), 379–502.

Benrabah, M. (2013). *Language conflict in Algeria: From colonialism to post-independence*. Multilingual Matters.

Bentahar, Z. (2020). "*Ytnahaw ga'!*": Algeria's cultural revolution and the role of language in the early stages of the spring 2019 Hirak. *Journal of African Cultural Studies, 33*(4), 471–488.

Benziane, I., & Boutaba, N. (2015). Waqie tatbiq nizam l.m.d fi aljamieat aljazayiria: dirasa maydania liwijhat nazar al'iidariiyn fi eayinat min aljamieat (The reality of implementing the LMD system in Algerian universities: A field study of administrators point of views in a sample of universities). *The Social and Human Sciences Review, 8*(2), 63–83.

Bessaid, O. N., & Benyahia, C. (2023). The role of e-learning in the Algerian Open University to achieve the development of human capital. In S. Sedkaoui, M. Khelfaoui, R. Benaichouba, & K. M. Belkebir (Eds.), *International conference on managing business through wen analytics* (pp. 167–178). Springer.

Bouchrika, I., Harrati, N., Wanick, V., & Wills, G. (2021). Exploring the impact of gamification on student engagement and involvement with e-learning systems. *Interactive Learning Environments, 29*(8), 1244–1257.

Bouherar, S., & Ghafsi, A. (2021). *Algerian languages in education: Conflicts and reconciliation*. Palgrave Macmillan.

British Council. (2014). *Algerians to study in UK*. https://opportunities-insight.britishcouncil.org/news/market-news/algerians-study-uk

British Council. (2021). Signing of the Memorandum – Higher Education and Scientific Research. *Online article*. https://www.britishcouncil.dz/en/programmes/education/higher-education-commission

Bude, U. (1983). The adaptation concept in British colonial education. *Comparative Education, 19*(3), 341–355.

Colonna, F. (2004). 'What has become of my friends?' Questions regarding the new Algerian 'diaspora' in France. *The Journal of North African Studies, 9*(2), 126–139.

D'amato, S., & Panta, G. D. (2017). Does rentierism have a conditional effect on violence? Regime oil dependency and civil war in Algeria. *The Extractive Industries and Society, 4*(2), 361–370.

Daghbouche, N. (2011). The impact of the credit transfer accumulation system on Algeria. *The Journal of North African Studies, 16*(3), 465–470.

Dreher, A., Lamla, M. J., Lein, S. M., & Somogyi, F. (2009). The impact of political leaders' profession and education on reforms. *Journal of Comparative Economics, 37*(1), 169–193.

Durham, B. (2020). *Une aventure sociale et humaine*: The Service *des Centres Sociaux* in Algeria, 1955–1962. In D. Matasci, M. B. Jerónimo, & H. G. Dores (Eds.), *Education and development in colonial and postcolonial Africa: Policies, paradigms, and entanglements, 1890s–1980s* (pp. 55–82). Palgrave Macmillan.

Faradji, M. A. (2015). 3rd World Conference on Psychology and Sociology, WCPS-2014: Sociological research in Algeria: Between theoretical language and social reality. *Procedia – Social and Behavioral Sciences, 185*, 352–356.

Garcia, N. (2014). Tensions between cultural and utilitarian dimensions of language: A comparative analysis of 'multilingual' education policies in France and Germany. *Current Issues in Language Planning, 16*(1–2), 43–59.

Gardinier, D. E. (1988). Les recommandations de la Conférence de Brazzaville sur les problèmes d'éducation. In I. Charles-de-Gaulle (Ed.), *Brazzaville, janvier-février 1944: aux sources de la décolonisation* (pp. 170–180). Plon.

Ghanem, D. (2022). *Understanding the persistence of competitive authoritarianism in Algeria*. Palgrave Macmillan.

Ghouati, A. (2022). Réformer l'enseignement supérieur par la professionnalisation des formations ? Comparaison de trois expériences en Algérie. In M. Benguerna & J. B. Meyer (Eds.), *L'Université et ses acteurs dans un monde en turbulence* (pp. 1–20). CREAD-CRASC.

Gradillas, M., & Thomas, L. D. W. (2023). Distinguishing digitization and digitalization: A systematic review and conceptual framework. *Journal of Product Innovation Management*. https://doi.org/10.1111/jpim.12690

Grandguillaume, G. (1997). Le Maghreb confronte a l' islamisme: Arabisation et demagogie en Algerie. *Le Monde Diplomatique*. https://www.monde-diplomatique.fr/1997/02/GRANDGUILLAUME/4593

Guessoum, A. (2019). Introducing quality assurance in Algerian higher education: The case of the University of Science and Technology Houari Boumediene. In A. Badran, E. Baydoun, & J. R. Hillman (Eds.), *Major challenges facing higher education in the Arab World: Quality assurance and relevance* (pp. 335–352). Springer.

Harbi, M. (2002). Le poids de l'histoire: Et la violence vint a l'Algerie. *Le Monde Diplomatique*. https://www.monde-diplomatique.fr/2002/07/HARBI/9218

Herreman, P. (1965). LES RELATIONS S'AGGRAVENT entre le régime Boumediene et le parti communiste français. *Le Monde*. https://www.lemonde.fr/archives/article/1965/09/13/les-relations-s-aggravent-entre-le-regime-boumediene-et-le-parti-communiste-francais_2201575_1819218.html

Holst, J. (2023). Towards coherence on sustainability in education: A systematic review of Whole Institution Approaches. *Sustainability Science, 18*, 1015–1030.

Idri, N. (2012). Education and reform to reach autonomous learners: Between reality and myth. *Procedia – Social and Behavioral Sciences, 69*, 2174–2183.

Kabel, A. (2023). The neoliberal linguistic consensus: Neoliberal multilingualism and linguistic governmentality in Morocco. *The Journal of North African Studies, 28*(4), 788–814.

Kadi, L. (2004). La politique linguistique Algerienne: Vers un etat des lieux. *Revue d'Amenagement Linguistique, 107*, 133–144.

Kadri, A. (2014a). Les conditions d'imposition du système scolaire français à l'Algéric. In A. Kadri (Ed.), *Instituteurs et enseignants en Algérie 1945–1975: Histoire et mémoires* (pp. 19–53). Karthala.

Kadri, A. (2014b). L'université d'Alger. In A. Kadri (Ed.), *Instituteurs et enseignants en Algérie 1945–1975: Histoire et mémoires* (pp. 175–185). Karthala.

Kessar, S., Rabab'Ah, G., Al-Khadra, W., & Hamdan, H. J. (2021). The representation of the Algerian Hirak protest movement in the international media: France 24 and Al-Jazeera. *Cogent Social Sciences, 7*(1), 1–18.

Khandriche, M. (1996). Algeria and France: The enduring strength of cultural ties. In C. Alden & J. P. Daloz (Eds.), *Paris, Pretoria and the African Continent: The international relations of states and societies in transition* (pp. 54–62). Palgrave Macmillan.

Khedouci, R. (2015). *Al madrasa wa al islah: 100 yawm fi al lajna al wataniya li islah al mandouma al terbawiya* [School and reform: 100 days in the national committee of educational reform]. Dar Alhadara.

Khelfaoui, H. (2003). Le champ universitaire Algérien entre pouvoirs politiques et champ économique. *Actes de la Recherche en Sciences Sociales, 3*(148), 34–46.

Kouachi, M., Sergui, J., & Rami, K. (2019). The reality of higher education in Algeria within the framework of the reform program. *Journal of Studies in Economics and Management, 2*(1), 299–316.
Laskaris, E. (2016). *La coopération culturelle franco-algérienne: Les coopérants français en Algérie indépendante: Enseignants des écoles (instituteurs et professeurs) et universitaires (1962–1980).* Doctoral thesis, Université Paris-Est. ProQuest Dissertations and Theses Global. https://theses.hal.science/tel-01691333
Layachi, A. (2001). The private sector in the Algerian economy: Obstacles and potentials for a productive role. *Mediterranean Politics, 6*(2), 29–50.
Le Roux, C. S. (2017). Language in education in Algeria: A historical vignette of a 'most severe' sociolinguistic problem. *Language & History, 60*(2), 112–128.
Little, J. P. (2005). *Une conquête morale: L'enseignement en A.O.F.* L'Harmattan.
Louanchi, T. (1969). L'Algérie expérimente l'enseignement programmé et Télévisé. *Audio-Visual Media, 3*(2), 21–27.
Mahdjoub, R., & Miliani, M. (2017). Education and career guidance in Algeria: Recurrent dysfunctions. In R. G. Sultana (Ed.), *Career guidance and livelihood planning across the Mediterranean: Challenging transitions in South Europe and the MENA region* (pp. 123–138). Sense Publishers.
Manière, L. (2010). La politique française pour l'adaptation de l'enseignement en Afrique après les indépendances (1958–1964). *Histoire de L'education, 128*, 163–190.
Maraf, B., & Vanci Osam, U. (2022). The smile revolution (hirak) as a driving force for an English 'tidal wave' and foreign language policy-making in Algeria. *Current Issues in Language Planning, 24*(2), 179–200.
Martin, I. (2003). Algeria's political economy (1999 2002): An economic solution to the crisis? *The Journal of North African Studies, 8*(2), 34–74.
Martinez, L. (2004). Why the violence in Algeria? *The Journal of North African Studies, 9*(2), 14–27.
Matasci, D., Jerónimo, M. B., & Dores, H. G. (Eds.). (2020). *Education and development in colonial and postcolonial Africa: Policies, paradigms, and entanglements, 1890s–1980s.* Cham: Palgrave Macmillan.
Mauviel, M. (2014). Le syndicalisme enseignant après l'indépendance. In A. Kadri (Ed.), *Instituteurs et enseignants en Algérie 1945–1975: Histoire et mémoires* (pp. 171–174). Karthala.
MESRS. (2021). Almutamar althaamin eashar liwuzara' altaelim aleali walbahth al ilmii fi alwatan ale arabi: nizam altaelim aleali walbahth al ilmii fi aljazayir, alwade alraahin wa efaq almustaqbal (The eighteenth conference of Ministers of higher education and scientific research in the Arab World: higher education and scientific research in Algeria, the current situation and future prospects) [*conference report*]. Algiers. https://alecso.org/cmae/algerie.pdf
MESRS. (2022a). *Grands agrégats* (Large aggregates). https://www.mesrs.dz/index.php/fr/agregats-2/
MESRS. (2022b). *Portail des plateformes numériques* (Digital platforms gate). https://www.mesrs.dz/index.php/fr/plateformes-mesrs/
Miliani, M. (2012). Teaching in higher education institutions in Algeria: A clash of paedagogies? *International Journal of Pedagogies and Learning, 7*(3), 218–226.
Mostari, H. A. (2004). A sociolinguistic perspective on Arabisation and language use in Algeria. *Language Problems and Language Planning, 28*(1), 25–43.

Musso, M. (2020). Oil companies as agents of post-colonial relations: France, Algeria, and Italy in the Sahara. In V. Dimier & S. Stockwell (Eds.), *The business of development in post-colonial Africa* (pp. 127–156). Palgrave Macmillan.

Naylor, P. C. (2000). *France and Algeria: A history of decolonization and transformation*. University Press of Florida.

Northey, J. A., & Guemar, L. N. (2020). The Algerian hirak: Youth mobilisation, elections and prospects for reform. *Economics and Culture of the Middle East, 2020*(II), 14–21.

Revere, R. B. (1971). *Consensus in independent Algeria, 1962–1965. PhD thesis.* Doctoral thesis, New York University. ProQuest Dissertations and Theses Global. https://www.proquest.com/docview/302435680/B28BF0FEEDB34AF4PQ/1?accountid=202267&sourcetype=Dissertations%20&%20Theses

Rezig, N. (2011). Teaching English in Algeria and educational reforms: An overview on the factors entailing students failure in learning foreign languages at university. *Procedia – Social and Behavioral Sciences, 29*, 1327–1333.

Rieckmann, M. (2018). Learning to transform the world: Key competencies in education for sustainable development. In A. Leicht, J. Heiss, & W. J. Byun (Eds.), *Issues and trends in education for sustainable development* (pp. 39–59). UNESCO.

Rouaba, A. (2022). Algerian leader in bold move to promote English at junior school. *BBC online article*. https://www.bbc.com/news/world-africa-62368931

Ruedy, J. (2005). *Modern Algeria: The origins and development of a nation* (2nd ed.). Indiana University Press.

Saulière, J. (2014). Corporate language: The blind spot of language policy? Reflections on France's Loi Toubon. *Current Issues in Language Planning, 15*(2), 220–235.

Schoelen, L. (2023). Between national identity, research and social function – Academics' perceptions of the ambivalent role of the Algerian university. In B. K. Daniel & R. Bisaso (Eds.), *Higher education in Sub-Saharan Africa in the 21st century* (pp. 221–233). Springer.

Silverstein, P. A. (2003). Martyrs and patriots: Ethnic, national and transnational dimensions of Kabyle politics. In J. McDougall (Ed.), *Nation, society and culture in North Africa* (pp. 85–109). Frank Cass.

Tillion, G. (1958). *Algeria: The realities* (R. Matthews, Trans.). Alfred A. Knopf.

Tilmatine, M. (2015). Arabisation and linguistic domination: Berber and Arabic in the North of Africa. In C. Stolz (Ed.), *Language empires in comparative perspective* (pp. 1–16). De Gruyter.

Uysal, H. H., Plakans, L., & Dembovskaya, S. (2008). English language spread in local contexts: Turkey, Latvia and France. *Current Issues in Language Planning, 8*(2), 192–207.

Weber, L. (2014). Louis Weber, entre syndicalisme et politique. In A. Kadri (Ed.), *Instituteurs et enseignants en Algérie 1945–1975: Histoire et mémoires* (pp. 249–260). Karthala.

Zartman, I. W. (1975). Algeria: A post-revolutionary elite. In F. Tachau (Ed.), *Political elites and political development in the Middle East: States and societies of the Third World* (pp. 255–292). Schenkman Publishing Company.

Zeraoulia, F. (2021). National reconciliation in Algeria from a bottom-up approach: Analysing victims' narratives. *The Journal of North African Studies, 7*(5), 862–893.

CHAPTER 4

Transformation to Digital Education: Beliefs and Practice

4.1 Introduction

In August 2023, the Minister of Higher Education Kamel Baddari proclaimed a significant milestone in the digital transformation of university registration for the academic year 2023/2024, heralding the advent of a paperless policy. This strategic shift aimed at not only alleviating the logistical challenges associated with physical attendance at universities but also mitigating the burdensome nature of traditional paperwork. The impetus for digitalisation was prominently featured in President Tebboune's December 2023 address to parliamentary representatives, wherein he outlined the imperative of comprehensive digital integration across sectors, particularly in education, as a means to combat corruption and enhance transparent administrative resource management. The Ministry, in adherence to this digitalisation agenda, implemented several digital platforms designed to alleviate the reliance on traditional paper-based processes within university services. This has notably facilitated newly recruited teachers in their ability to register online for teaching positions, eliminating the necessity of physically submitting voluminous paper files at university offices. Additionally, students studying abroad can now conveniently renew their scholarships through online channels, an option that was previously unavailable during the period from 2014 to at least 2020 when the onset of the COVID-19 pandemic obliged the Ministry to take online renewal seriously for abroad students. Furthermore, PhD candidates participating in contests in Algeria have transitioned to an electronic platform for submission, replacing the conventional practice of presenting a physical file to the university prior to the competition. Despite these commendable efforts and assertions by the official regime, an examination of the Algerian university landscape reveals a reality that diverges from the optimistic narratives. A notable challenge arises from the unreliability of certain digital platforms, with users frequently expressing discontent regarding frequent crashes. PROGRES, a platform crucial for student

S. Bouherar, S. Salem, *Sustainability in Algerian Higher Education*, https://doi.org/10.1007/978-3-031-81347-4_4

and teacher evaluation interaction, has been particularly susceptible to instability, attributed to the concurrent high electronic access by students and educators, resulting in systemic failures. This observation prompts a critical evaluation of the efficacy of the ongoing digitalisation initiatives within the Algerian academic sphere. The recurrent issues with platform reliability underscore the imperative for continuous refinement, technological fortification, and vigilance in ensuring the seamless integration of digital tools into academic processes. As Algeria pursues a digitally transformed educational landscape, addressing these challenges becomes paramount for realising the envisioned benefits of enhanced efficiency and transparency.

Despite the contrasting realities observed in university experiences and the narratives presented by the Ministry, the digitalisation of administrative activities should not be sensationalised in the media, nor should it make the headlines in the president's list of accomplishments. Often, the primary objective appears to be to embellish political achievements and assert the emergence of a new Algeria post-2019. As we approach the conclusion of President Tebboune's first term, these narratives persist as prominent highlights in his speeches. The underlying message is transparent: a potential re-election would signify a commitment to maintaining a policy of temporary solutions. This was explicitly conveyed by President Tebboune in his recent address to parliamentary representatives, specifically when discussing the unemployment grant for graduates who are yet to secure suitable employment. Proudly labelling it as "ça c'est des palliatives", he emphasised the significance of the unemployment grant in preserving the dignity of the nation's youth. The utilisation of tokenism policies in sustainable development (SD) also extends to the realm of palliative planning, often driven by political motives. The convergence of two seemingly disparate elements—SD and tokenism—hardly ever occurs under a single framework, but in the context of Algerian higher education (HE) practices and narratives, such convergence is evident.

In this chapter, our exploration emphasises the implications of digitalisation within the realm of HE, critically assessing its reliability for both students and educators. Our focus extends to scrutinising the outcomes and objectives stemming from the endorsement of digitalised HE, contending that it has arguably played a pivotal role in augmenting learning outcomes and teaching endeavours. A primary facet of our investigation involves aligning the intentions of digitalisation with SD outcomes. An essential factor contributing to this alignment is the role of research and its visibility in addressing local needs. Besides the prevailing assumption that digitalisation primarily fortifies the administrative aspects of universities with minimal emphasis on teaching and learning outcomes, our inquiry examines whether digitalisation has indeed contributed to elevating research centres, within those designated as National Research Programmes (PNR in French). This elevation aims to establish a connection between research outcomes and societal needs, thereby fostering a robust implementation of SD in HE. However, our investigation encounters contradictions, particularly in the realm of research funding. Despite the digitalisation of research plans, researchers continue to experience inadequate financial support for their projects, and more significantly, they receive no financial backing for publication. We begin by addressing Didham's (2018) claim, which posits that countries articulating robust policies for the integration of SD in education yet fall short in providing essential resources such as teacher training, materials, and research support are more likely to suffer from significant shortcomings of SD implementation. That is, such incongruities lead to an assertion that if a country's commitment to adopting SD in education lacks supportive infrastructure, the professed aspirations may be deemed contradictory. To substantiate our exploration, we analyse the findings of a comprehensive survey administered across 47 universities in 47 provinces of Algeria. This survey incorporates the perspectives of both teachers and students, shedding light on their experiences with the digitalisation of HE and e-learning. Through this empirical investigation, we aim to provide nuanced insights into the multifaceted dynamics of digitalisation's impact on HE in the Algerian context.

4.2 Key Characteristics of Digitalisation in Higher Education

The predominant top-down approach in Algerian HE stands out as the fundamental guiding principle, delineating the decision-making process from the highest echelons of power to the final implementation at universities. This hierarchical structure mandates that any education-related plan must secure approval from the supreme authority before trickling down through ministerial institutions, ultimately reaching universities. The recent imperative for digitalisation in HE, prompted by the exigencies of the global pandemic in early 2020, serves as a noteworthy example of this top-down governance. Following President Tebboune's directive to nationally close schools, colleges, and universities, the government was compelled to struggle with the challenge of preserving the academic year amidst the unprecedented disruption. This led to a swift contemplation of practical solutions, resulting in Algerian universities reaching out to educators between March and April 2020. The purpose was to encourage the dissemination of instructional content, including lessons, lectures, or personalised instructional videos, to facilitate student e-learning. The initiation of e-learning, however, unfolded amid a degree of confusion, as many educators lacked prior training on the utilisation of e-learning platforms. As the government recognised the imperative of salvaging the academic year around the end of 2020, it paved the way for a unique batching system in the academic year 2020/2021. This system divided a semester into two distinct periods, with each catering to specific cohorts of undergraduate and master's students. While this approach offered flexibility for universities to adapt the system according to available resources, it also introduced challenges, notably in the standards of learning. The implementation of face-to-face teaching alongside e-learning, with attendance not mandated in the former, resulted in an approximate eight-week teaching load for each batch in a semester. This blend posed challenges, prompting the government to find expedited solutions. The decision to allow undergraduate students to progress directly to master's studies without meeting previously stipulated criteria added another layer of complexity to the evolving educational landscape. By the end of 2021, the challenges were further exacerbated by the persistent pandemic and the government's lack of preparedness for e-learning, stemming from deficiencies in infrastructure and training. This unfolding scenario underscores the critical need for a re-evaluation of the top-down approach, particularly in the context of digitalisation, with an emphasis on fostering adaptability, autonomy, and comprehensive preparedness within the HE system.

A compelling argument emerges in favour of affording universities greater autonomy, positing that an increased level of institutional independence could yield more robust outcomes in the ongoing process of digitalising HE and enhancing e-learning experiences. The central role of governance becomes apparent in this context, as it serves as a key determinant in shaping the landscape of digitalisation, particularly in the realm of e-learning within HE. The influence of governance is perceptible in the formulation of centralised narratives and practices, largely dictated by the central government. The digitalisation of HE, particularly the implementation of e-learning platforms, has become a focal point where governance decisions have a profound impact. The centralised nature of decision-making can potentially hinder the diversity and adaptability needed for effective digital learning experiences. By fostering university autonomy, there exists the potential for a more decentralised and dynamic response to the challenges presented by the digital transformation of HE. An additional facet of e-learning worth considering is the pedagogical content, where guidelines for teaching and evaluation tend to exhibit uniformity across universities. While standardisation ensures a baseline of educational quality, it also raises questions about the adaptability and responsiveness of these guidelines to the diverse needs and contexts within individual institutions. The homogeneity in pedagogical approaches across universities may limit the exploration of innovative teaching methods that could cater to the unique learning styles and requirements of diverse student populations.

4.2.1 Governing E-learning

A prominent aspect of digitalisation lies in the centralisation of e-learning, marked by a uniform model implemented across various public universities and colleges nationwide. This standardised approach is characterised by the adoption of the Modular Object-Oriented Dynamic Learning Environment (Moodle) framework, which underscores the centrality of the "course" as the fundamental unit. Moodle places emphasis on the "course" as the core element which is made up of "sections" that include both "resources" and "activities" (Palumbo & Verga, 2015). Resources are content units provided to students within a course, while activities are interactive learning components designed for student engagement (Palumbo & Verga, 2015). The uniformity extends to the comprehensive form and functionalities of the e-learning platforms, ensuring consistency despite potential structural disparities among universities. Each teacher and student is provided with login credentials to access the platform, where lessons or lectures are presented in conventional formats such as Word and PDF documents. Furthermore, the platform facilitates the uploading of diverse educational materials, including books, research papers, blogs, videos, links, and files. Interactivity is fostered through chat options and assignment submissions, adding to the overall learning experience. It is noteworthy that these features, with minor variations, are ubiquitously integrated into every e-learning platform deployed across public universities in Algeria.

In contrast to the theoretically advocated learner-centred approach in e-learning experiences (An & Reigeluth, 2011), the e-learning platforms integrated into Algerian higher education institutions (HEIs) often exhibit a teacher-centred orientation. Knowledge typically flows from teachers, who act as the disseminators of information, to learners, who assume the role of recipients in the e-learning process. The stringent e-learning governance policies enforce a top-down structure, compelling teachers, students, and administrators to heavily rely on external social media platforms, such as Facebook, for the dissemination of crucial academic information. Notably, announcements regarding exams, academic issues, and other pertinent matters are frequently communicated through Facebook groups and pages, with limited visibility on the designated e-learning platform. This top-down approach reinforces authoritarian decision-making within the university, prioritising the institution's preferences over the needs and preferences of the university staff. An illustrative manifestation of this trend is the apparent low desirability among students to engage with e-learning platforms (Fatahi & Moradi, 2016). A more straightforward alternative is often embraced, supplanting the intricate functionalities of the official e-learning platforms. While the low desirability could be interpreted as a signal for the necessity of additional interventions for students (Fatahi & Moradi, 2016), there is a notable absence of comprehensive training programmes to acquaint students with the intricacies of the e-learning platform. This deficiency holds direct implications for SD in e-learning within HE contexts, as underscored by Zhou et al. (2020). The scarcity of training resources in Algeria is primarily attributed to the sheer volume of students, presenting a significant challenge in providing adequate instructional support to foster effective and sustainable e-learning practices. Addressing this gap in training resources is imperative to enhance student engagement, foster a more learner-centred environment, and promote the sustainable integration of e-learning in Algerian HEIs.

The governance mechanism in Algerian HE plays a pivotal role in shaping the landscape of e-learning, and it is essential to underscore its potential negative impact on the e-learning process. Goedegebuure and Hayden (2007) contend that a challenge in HE involves compliance with government policies, a dynamic influenced by the concept of trust stemming from the application of principles associated with good governance. However, the Algerian context diverges from the situation described by Goedegebuure and Hayden in some Western countries. While there is indeed compliance among HEIs, it is not grounded in trust but rather reflects an authoritarian ethos. This authoritarian inclination aligns with Foucault's (1991) concept of governmentality, wherein the government's desire for control is manifest in the provision of free public education to citizens. In the Algerian context, the government's approach to e-learning appears to mirror this authoritative stance. Rather than responding to the articulated needs of students, the government imposes its detached perspective on what e-learning should entail, diverging from the nuanced views and attitudes of students and teachers. This top-down governance model not only hinders the alignment of e-learning platforms with the actual needs of the academic community but also constrains the adaptability and responsiveness of these platforms to the evolving educational landscape. Consequently, there is a need for a more inclusive and collaborative approach to governance that considers the diverse perspectives and requirements of stakeholders in order to foster a more effective, student-centred, and flexible e-learning environment in Algerian HE.

4.2.2 *Standardising the Norms*

The examination of the e learning platform's superficiality and the ensuing standardisation opens a gateway to a more comprehensive analysis of the multifaceted implications that extend beyond the initial critique. As we navigate the convoluted landscape of Algerian HE, it becomes increasingly evident that the challenges posed by standardisation permeate various dimensions of the academic ecosystem. This prompts a thorough exploration, revealing the nuanced intricacies and underlying issues that warrant a more profound understanding. A collateral impact of the standardised approach is the encroachment upon the autonomy of universities to craft and deliver their e-learning content. The imposition of rigid formalities and ministerial decrees serves to shackle academic institutions, forcing them into a constrained framework dictated by state authorities. This rigid adherence stifles the potential for innovation, creativity, and diverse educational approaches that could cater to the unique needs of different disciplines and e-learning styles. The essence of HE, which traditionally thrives on intellectual diversity and exploration, is thus compromised by a standardised model that prioritises conformity over academic dynamism.

The repercussions of standardisation extend to the very core of student engagement in e-learning, revealing deeper concerns about the efficacy of the educational environment. While Ezeofor and Lent (2014) emphasise the role of supportive environments in fostering students' self-efficacy and positive outcomes, the existing discourse on e-learning in Algerian HE appears to overlook this critical aspect. The absence of attention to the supportive infrastructure required for students results in a deficiency in addressing their holistic development. Zalazar-Jaime et al.'s (2023) insights into the interplay between self-efficacy, active involvement, and academic satisfaction further accentuate the need for a comprehensive approach that prioritises the creation of an environment conducive to student success.

Teachers also are affected by the arising form of standardisation that pertains to the impediments it places on teachers in fostering peer interactions among students in e-learning platforms. In an era where collaborative learning and interpersonal skills are deemed crucial, the limited channels for promoting student engagement and interaction hinder teachers' efforts to develop these essential attributes (Zalazar-Jaime et al., 2023). Furthermore, the migration of both teachers and students to alternative platforms outside the university's purview underscores the shortcomings of the current e-learning paradigm. Platforms like Moodle and Udemy, offering greater flexibility and a diverse array of learning options, become attractive alternatives. This migration not only challenges the university's authority but also highlights the pressing need for a more adaptable and responsive e-learning infrastructure within the educational institutions themselves.

Another issue stemming from standardising the e-learning platforms is the assessment of e-learning content. Relying predominantly on project-based tasks for evaluation not only limits the scope of assessment methods but also raises questions about the credibility and fairness of such evaluations. The emphasis on group projects may inadvertently lead to concerns about individual contribution, thereby jeopardising the authenticity of the assessment process. Moreover, the shift towards project-based evaluations raises broader questions about the alignment of assessment strategies with the evolving demands of the job market. The emphasis on collaborative projects may align with certain industry needs but may not necessarily capture the diverse skill sets and individual capabilities of students.

The repercussions of standardisation extend further to the establishment of long-term strategic development goals within SD in HE. While the Ministry's narrative emphasises the digitalisation of HE, the disparity between administrative intentions and academic implementation signals a misalignment that requires urgent attention. The challenge lies in aligning the digitalisation narrative with the practical realities of academic implementation. The establishment of long-term SD goals demands a synergy between administrative visions and the actual needs of educators and students. A closer examination of the digitalisation narrative reveals a potential disconnect between administrative intentions and the intricacies of academic implementation. While there is a pronounced emphasis on digitalising HE, the actual translation of this vision into concrete academic plans appears to be lacking. The narrative, perhaps more applicable to administrative purposes, seems to overlook the intricacies of academic strategies, leaving a void in the alignment of digitalisation efforts with the core academic mission. A failure to bridge this gap may result in a digitalisation agenda that lacks the necessary academic depth and relevance, potentially rendering it incongruent with the broader educational objectives.

4.3 Methodology

In this chapter, we chose to employ an online survey approach, distributing it to 47 universities across Algeria, encompassing both teachers and students across various levels and majors. The decision to utilise a single questionnaire for all participants was based on the significant similarities observed in the e-learning systems and digital perspectives among the universities, all governed by a central ministerial entity dictating terms and decrees. The rationale behind employing a survey tool was twofold: the extensive participant pool and the goal of examining the satisfaction levels of both staff and students regarding the current objectives of digitalising HE that may not be fully captured by interviews or other methods that operate within smaller numbers. The inclusion of a large number of participants, spanning 47 provinces, and the standardisation of questions were key considerations in our choice of survey methodology. Additionally, given the uniformity in the e-learning systems and the centralised nature of regulations, a survey was deemed particularly apt for investigating attitudes towards digitalisation in HE. The adoption of subjective measures such as attributes and satisfaction, as suggested by Wilson (2013), further supported our decision to employ a survey as the preferred method for gathering insights into the perceptions and experiences of stakeholders in the digital transformation of HE.

To bolster participant engagement, we adopted a multi-pronged approach to augment the reach of the survey. Recognising the challenges in eliciting responses, especially in a landscape where time constraints often impede participation, we implemented strategies to emphasise the significance of participants' input. Personalised follow-up emails were created, emphasising the importance of teachers' perspectives in shaping the study's outcomes. We acknowledged the time constraints faced by educators and students, and therefore we highlighted the brevity of the survey and its potential impact on informing policies and practices in the realm of digital education. Moreover, recognising the pivotal role of interpersonal networks, we encouraged participants who had already taken the survey to share the link within their respective academic circles. This dissemination sought to tap into the inherent trust and camaraderie among colleagues and students, fostering a collaborative effort to ensure a more comprehensive dataset. In addition to the online outreach, we explored alternative channels, such as collaborating with university departments and educational associations using different social media platforms. This collaborative effort involved leveraging institutional support to endorse and promote the survey, emphasising its relevance to ongoing discussions on educational digitalisation. Through this collaborative network, we aimed to not only increase the participant pool but also to fortify the credibility of the study within academic circles. Acknowledging the diverse linguistic landscape in Algeria, efforts were made to offer language flexibility within the survey, accommodating both English and Arabic speakers. This linguistic inclusivity aimed to bridge potential communication gaps and ensure that participants could comfortably engage with the survey in their preferred language. While the response rate presented initial challenges, our commitment to obtaining a representative and diverse dataset propelled us to persist in our outreach efforts. We affirm that the personalised mailing list we generated comprised over 440 educators. Despite the substantial number, only a limited number of educators responded to the survey. Nevertheless, it is important to note that our survey encompassed nearly all state universities, spanning 47 provinces, with each province containing at least one public HEI.

Upon accessing the provided hyperlink, participants were initially presented with an information sheet available in two languages, English and Arabic. This document served as a comprehensive guide, elucidating the essence of the research and delineating the specific tasks required of the participants. Subsequently, participants were asked to endorse their voluntary participation by ticking a consent form box. The survey commenced with the initial question strategically designed to categorise participants into distinct groups based on their roles within the academic institution, distinguishing between teachers and students. The second inquiry delved into the demographic aspect, specifically targeting the age distribution of the participants. The third question sought to ascertain the geographical affiliation of participants, offering a selection among 58 provinces for a more nuanced understanding of the study's regional representation.[1] The fourth and fifth questions concentrated on the participants' access to essential technological resources, specifically probing the availability of personal computers and reliable internet networks for educational purposes. The ensuing set of questions, ranging from the sixth to the eighth including the tenth, meticulously probed the electronic infrastructure both within and outside the university context. This segment aimed to evaluate the internet reliability and technological facilities within the participants' respective academic institutions. The ninth question sought to discern the underlying motivation behind the digitalisation efforts, differentiating whether it primarily served educational or administrative purposes. Moving forward, questions 11–15 directed the participants' attention to the means of communication with the university and the specific parameters they believed would augment their e-learning experiences. This segment encapsulated the multifaceted aspects of communication channels and the key determinants contributing to an enriched e-learning environment for both educators and learners alike.

The analytical process relies on SPSS software for handling numerical data, predominantly originating from close-ended questions. Cronbach's alpha serves to determine the reliability of survey items (i.e. questions), encompassing various facets such as the reliability of internet use among students and staff and the availability of the internet at home, the university technological infrastructure and the maintenance of technical breakdown in the university, and the university technological infrastructure with the e-learning experience. This

[1] The total provinces of Algeria were 48 states but the government decided to include 10 more states in 2019, augmenting the number of provinces in Algeria to 58.

is important for assessing the coherence of responses across interconnected questions. Additionally, the application of the correlation test enables the examination of potential correlation between variables. In cases where open-ended questions are posed, thematic analysis is employed to uncover recurring patterns and prevalent themes within the responses provided across the three open-ended inquiries in the survey.

4.4 Reliability Test

In conducting the reliability test using SPSS, ten survey questions were incorporated as non-nominal (i.e. ordinal), while three nominal questions unsuitable for Cronbach's alpha calculation—pertaining to age, university name, and participant consent—were excluded from the analysis. Furthermore, open-ended questions underwent thematic analysis and were consequently omitted from the SPSS reliability test. The questions included when running the tests can be summarised as follows: the first question divided participants between teachers and students and we named it in the SPSS as (Position). The second question looked at participants' ownership of a personal computer, abbreviated as (PC). The third question asked about internet availability at home, referred to as (Internet). The fourth question asked about internet quality at home, abbreviated as (IQ). The fifth question asked about the availability of the university technological facilities, abbreviated as (UF). The sixth question was about assessing the maintenance of technological facilities in the university, referred to as (Maintenance). The seventh question asked about the purpose of digitalisation as informed by the Ministry of Higher Education, referred to as (Purpose). The eighth question investigated the commonly used devices for distance learning; the question is referred to as (Devices). The ninth question asked participants whether their e-learning experiences would qualify them to integrate globally in different e-learning experiences abroad in other developed countries; the question is referred to in the SPSS as (Integration). Finally, the tenth question asked about the university's most used means of communication to contact participants about their academic issues, abbreviated in the SPSS as (UniContact). The following two tables indicate the results of the reliability tests.

Table 4.1 presents Cronbach's alpha coefficients obtained from a sample comprising 404 participants drawn from diverse academic institutions. Cronbach's alpha, denoted by α, is a measure of internal consistency reliability, ranging between 0 and 1, with higher values indicating greater reliability. In

Table 4.1 Cronbach's alpha test

Cronbach's alpha	*Cronbach's alpha based on standardised items*	*N of items*
0.764	0.811	10

this study, the calculated Cronbach's alpha coefficient is $\alpha = 0.76$, signifying a positive and acceptable level of internal consistency among the survey items. This indicates that the items (i.e. questions) in the survey instrument are reliably measuring the underlying constructs, as evidenced by participants' consistent and coherent responses. Table 4.2 shows the mean and variance of the scale if the respective item is removed. It indicates the impact of each item on the overall mean and variability of the scale. Items with larger changes in mean or variance when deleted may have a greater influence on the overall scale score. The same table also represents the correlation between each item and the total score of the scale. It measures how well each item correlates with the overall construct (i.e. all relevant concepts that measure participants' attitudes about e-learning) being measured. Higher values indicate stronger relationships between the item and the underlying construct. The table also indicates the extent to which each item shares variance with the overall scale score. Higher squared multiple correlation values suggest that the item contributes more uniquely to the measurement of the construct. The last column in Table 4.2 shows Cronbach's alpha coefficient if the respective item is removed from the scale. Cronbach's alpha is a measure of internal consistency reliability, with higher values indicating greater reliability. Lower alpha values after item deletion may suggest that the item contributes positively to the overall reliability of the scale.

The consistently high values of Cronbach's alpha across all items, even when each item is deleted, indicate that the scale exhibits strong internal consistency reliability. This suggests that the items in the scale are closely related and measure the same underlying construct effectively. Specifically, the values range from 0.724 to 0.779, which are generally considered indicative of good internal consistency reliability. The slight variations in Cronbach's alpha coefficients when individual items are deleted suggest that each item contributes positively to the overall reliability of the scale. This indicates that none of the items are redundant or detract significantly from the scale's reliability when removed.

Table 4.2 Cronbach's alpha if item deleted

	Scale mean if item deleted	*Scale variance if item deleted*	*Corrected item-total correlation*	*Squared multiple correlation*	*Cronbach's alpha if item deleted*
Position	20.90	29.808	0.194	0.446	0.768
PC	20.58	27.951	0.613	0.603	0.743
Internet	20.52	27.857	0.653	0.530	0.741
IQ	19.29	23.831	0.560	0.507	0.724
UF	20.17	24.671	0.452	0.461	0.742
Maintenance	20.07	26.030	0.470	0.425	0.740
Purpose	21.13	27.484	0.532	0.403	0.742
Devices	18.62	23.431	0.337	0.514	0.779
Integration	19.38	23.196	0.511	0.540	0.733
UniContact	20.45	24.347	0.502	0.494	0.734

4.5 Normality Test

We performed a normality test on the sample of data derived from our survey to assess the adherence of this data to a Gaussian distribution. Subsequent to conducting the normality test using SPSS, Table 4.3 was generated, providing essential information for determining the appropriateness of employing parametric or non-parametric analytical techniques for investigating the survey questions. The small *p*-values (< 0.001) in the Sig. columns for both the Kolmogorov-Smirnov and Shapiro-Wilk tests suggest that the null hypothesis of normality, which assumes that the data are normally distributed, is rejected for all variables. These results indicate that the assumption of normality does not hold for the data. Therefore, when conducting statistical analyses, it is more appropriate to use non-parametric tests or robust methods that do not rely on the assumption of normality. One of the non-parametric tests we used as a result of Table 4.3 data is the Spearman test. Spearman's rho is a valuable alternative for assessing relationships between variables. This is particularly relevant when conducting correlation analyses where the assumption of normality does not apply to the population. By using Spearman's rho, we can determine whether there is a statistically significant monotonic relationship between the variables without relying on normality assumptions. The Spearman correlation coefficient ranges from -1 to 1, where a value close to 1 indicates a strong positive monotonic relationship, a value close to -1 indicates a strong negative monotonic relationship, and a value close to 0 indicates no monotonic relationship.

Table 4.3 Normality test

	Kolmogorov-Smirnov			*Shapiro-Wilk*		
	Statistic	*df*	*Sig.*	*Statistic*	*df*	*Sig.*
Position	0.369	404	< 0.001	0.632	404	< 0.001
PC	0.455	404	< 0.001	0.537	404	< 0.001
Internet	0.467	404	< 0.001	0.466	404	< 0.001
IQ	0.173	404	< 0.001	0.917	404	< 0.001
UF	0.262	404	< 0.001	0.840	404	< 0.001
Maintenance	0.288	404	< 0.001	0.815	404	< 0.001
Purpose	0.449	404	< 0.001	0.326	404	< 0.001
Devices	0.354	404	< 0.001	0.714	404	< 0.001
Integration	0.163	404	< 0.001	0.885	404	< 0.001
UniContact	0.329	404	< 0.001	0.604	404	< 0.001

4.6 The Relationship Between Participants' Academic Position and PC Ownership

Table 4.4 provides data about the relationship between the participants' academic position and their ability to own a PC in the academic setting of their institutions. Since participants are both teachers and students in HE, we anticipated a more reliable relationship between teachers that are coded in the SPSS data as 2 and PC ownership that is coded in the SPSS data as 2.

The correlation coefficient between PC ownership among participants (teachers and students) is 0.352. The asterisks next to the coefficient (0.352) typically indicate the level of statistical significance. In this case, a significance level of $p < 0.001$ suggests a statistically significant positive correlation between PC ownership among participants. This suggests that the correlation between PC ownership and participant positions is unlikely due to random chance. However, the correlation coefficient of 0.352 suggests a moderate positive correlation between PC ownership and participant positions (teachers vs students). This implies that as the ownership of personal computers increases among participants, the likelihood of them being teachers also tends to increase and vice versa.

Teachers, often in more stable employment positions, are more likely to own PCs, which they may use extensively for academic research, lesson planning, grading, and communication with students. On the other hand, students, particularly those from lower-income backgrounds, may face financial constraints that limit their ability to own personal computers solely for academic use. This can potentially affect their access to educational resources and opportunities for online learning. Linking these findings to SD objectives highlights the importance of equitable access to technology in achieving educational goals. Sustainable Development Goal 4 (SDG4) aims to ensure inclusive and equitable quality education and promote lifelong learning opportunities for all, as discussed in Chap. 2. However, disparities in access to PCs and other technological resources can hinder progress towards this goal by perpetuating inequality in educational outcomes. Addressing these disparities requires targeted interventions to improve access to technology for marginalised groups, including students from low-income backgrounds. Also, the observation that many students and even some faculty members may not own personal

Table 4.4 Position and PC correlation

			Position	*PC*
Spearman's rho	Position	Correlation Coefficient	1.000	0.352
		Sig. (2-tailed)	–	< 0.001
		N	404	404
	PC	Correlation Coefficient	0.352	1.000
		Sig. (2-tailed)	< 0.001	–
		N	404	404

computers for academic purposes also raises questions about resource allocation within universities. If a significant portion of the academic community lacks access to essential technological resources, it may indicate a need for the university (or perhaps the government) to invest in infrastructure and support services to ensure equitable access to technology. This could include initiatives such as providing loaner laptops, establishing computer labs with free access, or offering subsidies for purchasing computers for academic use.

4.7 Internet Quality

Table 4.5 shows data concerning participants' perceptions regarding the provision and quality of internet services within their domiciles. Specifically, the focus is on gauging the dependability and calibre of the internet exclusively within the home environment. This inquiry delves into participants' assessments of internet quality at home and the attendant appraisal of its reliability across diverse functional domains, including academic pursuits. The objective is to scrutinise the extent of infrastructure accessibility facilitated by governmental provisions to its populace, given the university's prior elucidation of intermittent internet provisioning for its academic constituents.

The moderately positive correlation (rho = 0.451) observed between internet quality and internet usage highlights a noteworthy association: as internet quality improves, there tends to be an increase in internet usage levels. However, this relationship does not exhibit the strength one might expect, given assumptions about the widespread accessibility and quality of the internet in Algeria, the context of this study. This finding suggests that while higher internet quality does indeed encourage greater internet usage, there are factors beyond quality alone influencing internet utilisation. Within the academic community, which constitutes the sample of this study, this correlation holds particular significance. It implies that scholars, researchers, and students are responsive to improvements in internet quality, likely leveraging it for various academic pursuits such as research, collaboration, and accessing educational resources. For policymakers and governmental bodies, this result carries important implications. It suggests that initiatives aimed at increasing internet usage, particularly within academic circles, should prioritise enhancements in internet quality. By investing in infrastructure and measures to bolster internet quality, the

Table 4.5 Internet quality and use

			Internet	*IQ*
Spearman's rho	Internet	Correlation coefficient	1.000	0.451
		Sig. (2-tailed)	–	< 0.001
		N	404	404
	IQ	Correlation coefficient	0.451	1.000
		Sig. (2-tailed)	< 0.001	–
		N	404	404

government can potentially stimulate greater engagement with online resources and foster a more digitally literate population. Furthermore, to bolster the significance of this finding, it is worth noting that despite a reported 81% of participants claiming internet availability and use at home, only 38.5% perceive the quality of their home internet as reliable and of good quality. This shows a disconnect between mere accessibility and the perceived adequacy of internet services, further emphasising the importance of prioritising internet quality improvements alongside efforts to expand access.

The digital landscape between practices and beliefs reveals notable disparities. Particularly within university settings, while e-learning garners urgent attention, the absence of requisite digital infrastructure hampers the effective implementation of transformative teaching methodologies. Similarly, governmental initiatives aimed at elevating digitalisation in HE face scrutiny due to inadequate infrastructure, undercutting decision-making processes. Challenges persist in internet accessibility and quality, posing hurdles to the administration's objectives. These concerns are underscored by data from Algerie Telecom's website, which highlights that only 5.5 million individuals, including a mere 1.1 million optical fibre subscribers, out of a population exceeding 42 million, access internet services at home (Algerie Telecom, 2024). This highlights the urgent need for comprehensive infrastructural development and strategic interventions to bridge the digital gap and realise the full potential of digitalisation in education and beyond.

4.8 University Facilities and Maintenance

Table 4.6 illustrates the responses of participants regarding the accessibility of university facilities and the maintenance services provided by the institution in the event of technical malfunctions affecting teaching amenities. In other words, in this table, we analyse the availability and functionality of essential educational facilities in the academic community.

The Spearman's rho correlation coefficient between UF and maintenance is found to be 0.549, which indicates a positive correlation between these two variables, but not a very strong one. This suggests that as the quality or availability of university facilities increases, there tends to be a corresponding increase in the effectiveness or adequacy of maintenance services provided by

Table 4.6 Higher education technology infrastructure

			UF	*Maintenance*
Spearman's rho	UF	Correlation coefficient	1.000	0.549
		Sig. (2-tailed)	–	< 0.001
		N	404	404
	Maintenance	Correlation coefficient	0.549	1.000
		Sig. (2-tailed)	< 0.001	–
		N	404	404

the institution. Similarly, when maintenance services are more efficient or comprehensive, there is likely to be a positive impact on the overall condition and functionality of university facilities. Furthermore, the significance level associated with this correlation coefficient is reported to be less than 0.001 ($p < 0.001$), indicating a highly significant relationship between UF and maintenance. This implies that the observed correlation is unlikely to have occurred randomly and provides evidence to support the existence of a meaningful association between these variables within the context of the study.

The examination of the availability of facilities is thoroughly addressed in Sect. 4.12.2 of this chapter. It is imperative to note that an assumption, later substantiated in Sect. 4.12.2, posits that a considerable number of HEIs, particularly those administered by the Ministry of Higher Education, suffer from a pronounced deficit in technological facilities designed for educational purposes. What emerges as a novel variable in this equation is the decline in maintenance services, a consequence attributed to either the inadequate availability and functionality of facilities or the insufficient qualifications of technicians overseeing these facilities. Regardless of the specific causative factor, the persistent nature of this issue within the academic landscape gives rise to grievances among both educators and students. In fact, more than 70% of the study participants expressed dissent regarding the assumption that their respective institutions furnish them with commendable technological facilities tailored for teaching and learning purposes. This dissent emphasises a prevalent dissatisfaction with the existing technological infrastructure in place. Additionally, a significant majority, exceeding 66% of participants, manifests unfavourable sentiments regarding the maintenance of these facilities, assigning a poor rating to the overall upkeep. This collective discontent further accentuates the gravity of the challenge, indicating that not only is the availability of facilities a concern, but the subsequent maintenance practices significantly contribute to the perceived inadequacy of the technological support system.

4.9 The Devices Used in E-learning

In this section, we explore the relationship between PCs and the devices utilised by individuals engaged in e-learning activities. However, in Sect. 4.6, we explored the association between PC ownership and academic roles, concluding that educators are more inclined to possess a PC compared to students. Table 4.7 supplements this discussion by presenting additional insights into the devices predominantly utilised by participants in their remote learning endeavours. This variable provides a spectrum of options for participants, including smartphones, tablets (iPads), PCs, desktop computers, and even no device at all. Among these choices, PCs and smartphones emerge as the most favoured. Notably, PCs exhibit the highest utilisation rate, comprising 56.4% of all participants, while smartphones account for only 32.7%.

The correlation coefficient between PCs and other devices is 0.632, indicating a good positive correlation between the usage of PCs and the adoption of

Table 4.7 Devices used by participants in distance learning

			PC	*Devices*
Spearman's rho	PC	Correlation coefficient	1.000	0.632
		Sig. (2-tailed)	–	< 0.001
		N	404	404
	Devices	Correlation coefficient	0.632	1.000
		Sig. (2-tailed)	< 0.001	–
		N	404	404

other devices by participants in distance learning. The significance level is denoted by "< 0.001", indicating a *p*-value of less than 0.001, implying a highly significant relationship. A correlation coefficient of 0.632 between PCs and other devices suggests that as the usage of PCs increases among participants in distance learning, there tends to be a corresponding increase in the adoption of other devices. This indicates that individuals who use PCs for distance learning often may also rely on additional devices, such as smartphones, potentially for supplementary learning activities or accessibility purposes. The findings also imply a direct relationship between the prevalence of PC ownership among participants and its utilisation in e-learning, especially when considering that 76% of participants claimed ownership of a PC. This suggests that as the ownership of PCs increases within the participant group, there is a corresponding escalation in their engagement with e-learning. Importantly, these results challenge the notion that PC usage is exclusive to teachers, highlighting that students are active contributors to the increased utilisation of PCs in the context of e-learning.

The survey data indicating that 76% of participants claim ownership of PCs while 32.7% utilise smartphones for distance learning suggests a nuanced relationship between devices in educational contexts. On the one hand, it implies that alternative devices, such as smartphones, serve either as supplements to PC usage or as substitutes in certain learning scenarios. This emphasises the adaptability of learners and the diverse ways in which technology is integrated into education. However, beneath this technological diversity lies a deeper issue of digital equity and inclusion within the Algerian HE landscape. The coexistence of high PC ownership alongside significant smartphone usage points to disparities in access to technology and potentially unequal opportunities for participation in e-learning initiatives. This echoes concerns raised by Grimes and Porter (2023) and Schmidt-Crawford et al. (2023), regarding the lack of digital equity and inclusion in similar educational contexts. Furthermore, this section of the analysis serves to validate and contextualise the findings discussed in previous sections, notably Sects. 4.7 and 4.8, as well as Sect. 4.12.2. It reinforces the significance of understanding the multifaceted dynamics of technology adoption in educational settings and highlights the need for targeted interventions to promote digital equity and inclusive learning environments.

4.10 Participants' Means of Communication with Their Institutions

Table 4.8 elucidates the correlation coefficient between the university's modes of contact with its academic society, as represented by UniContact, the purpose of digitalisation, and the academic position of participants. UniContact is categorised into three choices: Facebook (coded as 1), e-learning platforms (coded as 2), and emails (coded as 3). Simultaneously, the Purpose variable gauges participants' perspectives on digitalisation intentions aligned with either administrative purposes (coded as 1) or teaching purposes (coded as 2), mirroring the ministry's intentions. The position of participants is divided between teachers (coded as 2) and students (coded as 1). Significantly, over 54% of participants claim Facebook as the primary communication tool with their institutions. Furthermore, a substantial majority, exceeding 81%, perceive the purpose of digitalisation as primarily serving administrative goals rather than teaching objectives. These statistics provide valuable context to interpret the findings in Table 4.8, shedding light on the preferences and perceptions of participants regarding the intersection of university contact methods and the digitalisation agenda.

The correlation coefficient between Purpose and UniContact is −0.035. This value suggests a very weak negative correlation between UniContact and Purpose overall. The *p*-value (Sig.) associated with the correlation coefficient between Purpose and UniContact is 0.486, indicating that the correlation is not statistically significant. The lack of a significant correlation implies that the choice of means of contact (UniContact) within the university setting, whether it be Facebook, e-learning platforms, or emails, does not reliably predict whether digitalisation efforts are geared more towards teaching or administrative purposes. The data presented, however, indicates that a significant portion of participants, specifically over half, perceive Facebook as a primary communication tool with their institutions. Moreover, a substantial majority still associate digitalisation primarily with enhancing administrative processes. These statistics emphasise a glaring deficiency in prioritising digital e-learning

Table 4.8 Means of communication between participants and their institutions

			Purpose	*UniContact*	*Position*
Spearman's rho	Purpose	Correlation Coefficient	1.000	−0.035	−0.164
		Sig. (2-tailed)	–	0.486	< 0.001
		N	404	404	404
	UniContact	Correlation Coefficient	−0.035	1.000	0.567
		Sig. (2-tailed)	0.486	–	< 0.001
		N	404	404	404
	Position	Correlation Coefficient	−0.164	0.567	1.000
		Sig. (2-tailed)	<0.001	<0.001	–
		N	404	404	404

initiatives, especially considering the substantial reliance of HE on social media platforms.

The correlation coefficient between UniContact and Position can provide more insights into the discussion of digitalisation in Algerian HE. There is a strong positive correlation between the modes of communication, including various channels such as Facebook, e-learning platforms, and email, and the position of participants. The correlation coefficient of 0.567 suggests that as the mode of communication increases (particularly emails), the likelihood of the respondent being a teacher also increases. The p-value being less than 0.001 indicates that this correlation is statistically significant. This finding implies that teachers are more inclined to use various means of communication, including e-learning platforms and email, compared to students. The strong correlation suggests that the highest mode of communication (email) is closely associated with the position of the respondents as teachers. While the correlation suggests a higher likelihood for teachers to utilise email as a means of communication compared to students, it is important to note that not all participants necessarily rely solely on emails for communication with their institutions. The expectation is that teachers, being more attuned to digitalisation, may demonstrate greater responsiveness to such communication mediums than students.

4.11 Participants' Attitudes Towards the Quality of Facilities in Higher Education

Table 4.9 represents the correlation coefficient between University Facilities (UF), Integration, and Maintenance variables. We have previously discussed UF and Maintenance variables in Sect. 4.8 under Table 4.6. However, our current focus is on examining the correlation coefficient specifically between UF and Integration, as well as Maintenance and Integration. In other words, we compare the Algerian e-learning scenario with other developed countries, particularly in terms of how participants perceive the ease of integrating into e-learning contexts abroad and their satisfaction levels regarding facilities provision, including University Facilities and Maintenance.

Table 4.9 Participants' satisfaction with higher education e-learning facilities

			UF	*Integration*	*Maintenance*
Spearman's rho	UF	Correlation coefficient	1.000	0.432	0.549
		Sig. (2-tailed)	–	< 0.001	< 0.001
		N	404	404	404
	Integration	Correlation coefficient	0.432	1.000	0.368
		Sig. (2-tailed)	< 0.001	–	< 0.001
		N	404	404	404
	Maintenance	Correlation coefficient	0.549	0.368	1.000
		Sig. (2-tailed)	< 0.001	< 0.001	–
		N	404	404	404

The correlation coefficient of 0.432 between UF and Integration indicates a moderate positive correlation. This suggests that participants who perceive better UF also tend to find it easier to integrate into e-learning environments abroad. The statistical significance, with a significance level (Sig.) of <0.001, underscores a good correlation. However, despite its statistical significance, the correlation is not exceptionally strong. This implies that there may be room for improvement in participants' satisfaction with the current facilities provided for e-learning in Algerian universities. Indeed, it is noteworthy that over 70% of participants express disagreement regarding the status of e-learning facilities provision nationwide. This finding reveals the importance of enhancing e-learning facilities to better meet participants' needs and expectations. It becomes evident that greater satisfaction with the current e-learning infrastructure correlates with improved integration into more sophisticated e-learning environments abroad. Therefore, prioritising enhancements in e-learning facilities can significantly contribute to fostering a seamless transition and integration for Algerian participants into global e-learning landscapes. This observation aligns closely with the third target of the fourth goal outlined by the United Nations, as elaborated in Chap. 2, Sect. 2.3. The third target emphasises achieving equal access to technical, vocational education and training (TVET), along with HE opportunities. An essential means in addressing this target is ensuring the provision of safe and inclusive learning environments, as advocated by UNESCO (2018). As a result, by enhancing e-learning facilities and fostering a more satisfactory environment for education, particularly in Algerian universities, we are actively contributing to the realisation of this UN goal. Safe and inclusive learning environments are pivotal not only for equitable access but also for facilitating effective integration into global educational landscapes, thus empowering individuals to thrive in diverse educational settings. This would show the interconnectedness between local efforts to improve educational infrastructure and broader international initiatives aimed at advancing educational equity and inclusivity.

The correlation coefficient of 0.368 between Integration and Maintenance signifies a moderate positive correlation, with a statistically significant level (Sig.) of < 0.001. This suggests that participants who experience greater ease in integrating into e-learning contexts abroad also tend to perceive better maintenance and vice versa. As we explored in Sect. 4.8, Maintenance is intricately linked with UF, with the provision of better UF often resulting in improved Maintenance standards within universities. Therefore, the correlation between Integration and Maintenance reinforces the interconnected nature of these variables, indicating that positive experiences with Integration are likely to correspond with positive perceptions of Maintenance. However, it is important to note that the correlation between Integration and Maintenance is not particularly strong. This implies that while there is a relationship between these factors, it is not deterministic; positive aspects (other than Maintenance) of integration do not necessarily guarantee a positive reaction towards maintenance. This underscores the need for universities to address integration (as a potential global feature of Algerian e-learning) comprehensively to ensure a conducive and supportive e-learning environment for all participants. The following section employs thematic analysis to examine the remaining open-ended questions.

4.12 "I Do Not Think the University Has Good Higher Education Learning Quality"

The title of this section denotes a participant's reaction to their discontent with e-learning, serving as an organising theme capturing responses from participants across three open-ended survey questions. This theme has given rise to five fundamental sub-themes (basic themes), each intricately linked to participants' dissatisfaction with e-learning experiences. Despite a keen interest in investigating the influence of digitalisation on HE in Algeria, it is observed that digital higher education has prioritised enhancing administrative aspects over pedagogical ones. Consequently, the exclusive manifestation of digitalisation in pedagogy is indicated through e-learning, as elucidated earlier in this chapter.

One of the emerging issues in students' dissatisfaction concerns the attitudes of both students and educators towards the adoption of e-learning. It appears that certain educators and a majority of students lack adequate training for engaging with e-learning, both before and after the onset of the COVID-19 pandemic. This lack of readiness to utilise e-learning platforms for various purposes has prompted administrators, faculty, and students to turn to social media as an alternative means of communication within the academic community (i.e. teachers and students). Survey findings indicate that over 50% of educators and students across 47 universities nationwide rely on Facebook as the primary communication channel between themselves and their respective institutions, while only 4.5% utilise e-learning platforms for the same purpose. It could be contended that such platforms are primarily utilised for the dissemination of pedagogical content to students, thus validating various perspectives on this issue. Firstly, there exists no standardised mode of communication between the university and its academic community, as e-learning platforms predominantly serve as repositories for class materials, with minimal provision for interactive communication between administrators and members or between educators and students. Given the large student populations, universities often opt for more inclusive methods such as social media, which require no specialised training or provision of login credentials by the institution. Additionally, social media ensures widespread dissemination of information and enjoys higher visibility compared to e-learning platforms, which are perceived as less reliable and visible to the university community.

Moreover, educators have a more reliable means of communication with their respective institutions compared to students. The survey reveals that only 40% of the sampled individuals (predominantly educators) utilise email as their primary method of contact with universities. This further underscores the lack of a sophisticated technological infrastructure for e-learning, wherein learning, assignment submission, evaluation, and communication are seamlessly integrated within the system. Unlike the blackboard technology utilised in universities in the UK or USA (referred to interchangeably as Moodle, UniHub, Student Portal, Dashboard, etc.), Algerian university e-learning systems do not offer a comparable platform. The current e-learning system in Algerian universities primarily facilitates the uploading of lesson materials online, without robust features for interactive communication. In contrast, Facebook provides similar content with enhanced interactive capabilities among users and fewer concerns regarding technical failures. This explains why both students and educators resort to Facebook as an alternative to existing e-learning platforms. According to a study conducted by Sayaf (2023), students' satisfaction levels in e-learning can be influenced by five key elements: peer interaction, interactive lectures, information quality, system quality, and service quality. Incorporating these criteria into the e-learning process would have direct positive implications for the long-term SD plans in HE. The dissatisfaction expressed by the university community indicates the absence of these criteria in the current e-learning environment within Algerian universities. While these criteria contribute to enhancing students' satisfaction with e-learning, our research suggests the existence of relatively distinct features, which will be discussed in the subsequent sections.

4.12.1 "E-learning Platforms Must Be Fully Functional Not Just a Space to Upload Documents"

This section explores participants' critiques regarding the effectiveness of Algerian e-learning platforms. Expanding on the chapter's earlier discussion of the standardised, one-model-fits-all approach to e-learning in Algerian universities, the participants contend that this rigid uniformity negatively impacts the overall effectiveness of online learning initiatives. They argue that the inflexibility inherent in this system stifles the potential for adaptation and tailoring to the specific needs and contexts of individual universities due to "central government" dictations. Additionally, the resources offered on these platforms reflect a blind strategy of directly uploading the exact content of lessons online, potentially providing an opportunity for students who frequently miss classes to access the lesson content and thus assist non-attending students in catching up. This potentially undermines academic integrity and the seriousness of pursuing academic achievement. Furthermore, the issue of student absences in universities has become less of a concern for the administration, a trend that was solidified during the early stages of the COVID-19 pandemic in 2020 in Algeria. This trend was formalised through the Ministerial Decree No. 633 issued on 26 August 2020, by the Minister of Higher Education. The decree stipulates that students' attendance in lectures or tutorials (Travaux Dirigés, or TD, in French) is not mandatory, evaluations of students can be conducted either face to face or through distance learning, students can take exceptional study leave due to illness or fatigue, and the criteria for progressing from one university level to another have been significantly relaxed to be more accessible to students regardless of their academic achievements. Notably, the decree also mandates that every teacher must upload their lessons and lectures onto the designated e-learning platform.

The participants' dissatisfaction also highlights significant concerns regarding the quality of Algerian e-learning platforms. These concerns stem from a critique of the uninspired deployment of resources and communication spaces that these platforms lack. One participant's response to that is the following:

> It cannot be called distance learning because the teacher only puts some of the lessons written on the podium. This is no different than putting lessons in social networking platforms or sending them directly through e-mail to students. Based on personal experience, the distance learning platform does not have visible recorded lessons, and the exact space for uploading lessons on the platform is very few, estimated at only B2 Mega, cannot be used as a reference for students. Also, students can only benefit from interactive exercises on the platform using a special account and most students do not have an account on the distance learning platform and sometimes face multiple problems of accessing it because of weak Internet or through a technician on the platform.

During the peak of the COVID-19 pandemic, the utilisation and creation of such platforms surged, aiming to swiftly address the closure of schools and universities by offering easy access to previously face-to-face lessons. However, a notable issue with these platforms is the absence of interactive spaces for students and teachers. "I think the university should have interactive online platforms, engaging multimedia content, and accessible support services to enhance the e-learning experience", claims one participant. Another participant advocates for the creation of a platform adhering to international standards to "improve digital proficiency and embrace the power of communication". Furthermore, concerns about the quality and effectiveness of these platforms are prevalent among participants. They stress the importance of building virtual communities, establishing external partnerships for diverse resources, and continuously improving platforms based on user feedback. "Building virtual communities, exploring external partnerships for diverse resources, and continuous improvement based on user feedback would be key priorities", claims one participant. These statements underscore the challenges regarding the quality and efficiency of e-learning platforms, which often lack robust features that reflect authentic e-learning experiences. A significant concern highlighting the subpar quality of these platforms is the inadequate provision of resources to learners. Students frequently express dissatisfaction with the lack of "accessible resources" beyond their lessons. Notably, there is a scarcity of resources that students can access or use for free, aside from their course materials. Additionally, students and teachers are unable to access paid resources using their institution credentials, resulting in platforms predominantly featuring content produced by teachers for students.

4.12.2 "I Always Complain About the Unavailability of the Necessary Equipment in the Classrooms and the Lack of Internet"

This thematic concern encapsulates one of the educators' dissatisfaction with the limited provision of internet connectivity within the university campus, primarily restricted to administrative functions. Harmon (2007) articulates various justifications for educators across colleges to impart internet information literacy skills to their students, emphasising the cultivation of critical thinking within the academic setting. Notably, Trend et al. (1999) characterise this situation as the reality gap, elucidating the disparities between real educational practices and government rhetoric. Despite governmental assertions, the reality remains that Algerian universities grapple with a profound deficit in basic internet infrastructure. The implementation of an e-learning platform, while ostensibly addressing educational needs, fails to ensure universal internet access for students within the university premises. A teacher contends that students "cannot attend their online courses because of the lack of internet access" and highlights the challenges faced by instructors in conducting evaluations, particularly in an online format. The absence of internet connectivity within the campus adversely impacts both pedagogical processes and students' academic performance, especially when reliance on the e-learning platform for assessments is sometimes imperative. The repercussions extend beyond academic realms, imposing financial burdens on students who must manage their testing, evaluation, and online learning expenses. A student voices the struggles, stating, "Personally, I am from another Wilaya (province) and I am obliged to spend more than 2GB data per month for getting access to lectures". Notably, the university does not cover students' internet or computer expenses, further exacerbating challenges for those from low-income families and potentially hindering their ability to engage in any form of online learning within the university.

The absence of internet access within the academic community of the university poses a challenge for students in developing Information and Communications Technology (ICT) skills. If Harmon (2007) enumerates rationales for integrating internet information literacy skills into the curriculum, the dearth of internet connectivity can be identified as the primary impediment to incorporating such skills. A teacher laments, "I wish we could organise regular training days, each time covering only one aspect with practice", highlighting the inadequacy of ICT training. This insufficiency in training correlates closely with the aforementioned lack of facilities and internet access. Expressing the need for comprehensive support, one participant asserts, "I think it is important to provide both facilities and equipment including laboratories and libraries equipped with computers and access to the internet for students, both to motivate them and enhance the e-learning experience". The deficiency in ICT training and internet literacy skills may consequently stem from the failure to "rethink infrastructure" within the university's technological facilities. The absence of internet access and infrastructure facilities logically contribute to the prevalence of high levels of internet information illiteracy among both students and teachers, thereby affecting the attainment of SD expectations in HE.

4.12.3 *"Training Programmes for both Teachers and Students on Using Moodle and PROGRES Platforms"*

A prominent concern among participants regarding e-learning pertains to the challenge of adequately training both teachers and students to use e-learning platforms. Despite the purported simplicity and fundamental nature of these platforms, the pervasive issue of technological unfamiliarity persists within Algerian universities. The abrupt integration of e-learning into HE left little room for meticulous planning by university administrations. In the lead-up to the implementation of e-learning initiatives, such as in early 2020 at institutions like Bordj Bou Arreridj University, efforts were made to address technical hurdles. For instance, this university appointed e-learning technicians tasked with uploading teaching materials, managing teacher accounts, and providing technical support as needed. In contrast, other universities opted to encourage instructors to independently upload content onto the platforms. However, despite the provision of limited training sessions, these initiatives often lacked mandatory participation from educators, with attendance being merely optional.

The challenge extends to students as well, who frequently express dissatisfaction with the lack of preparation in utilising these digital platforms. According to one participant, "most teachers and students have found it difficult to digitally adapt, that is, to use technology". Another participant suggests the need to "organise sessions to confirm that all the students are able to access the internet". The deficient internet information literacy skills among students hinder their capacity to engage with e-learning platforms. This issue is exacerbated by the university's inadequate technological infrastructure and limited internet accessibility, as previously discussed. Given the unfavourable feedback regarding students' hesitance towards utilising e-learning platforms, it is reasonable to assume that the university should have been aware of the technical challenges and the lack of training, particularly for students. This may elucidate the university's decision to utilise social media platforms as alternative means to promote various educational activities and events, including exams, timetables, make-up sessions, and academic gatherings such as conferences and seminars. One participant recounts missing make-up exams because timetable announcements were exclusively made on the department's Facebook page, which they could not access as they do not have a Facebook account. Another participant criticises the reliance on Facebook for disseminating such information, stating, "There is no point to celebrate the policy of digitalisation whilst teachers have to write quizzes and make-up session announcements on papers then take pictures with phones to be posted by the head of department on Facebook". This participant further emphasises, "This is really an old-fashioned way of communication that should not be used in academia in the twenty-first century!" Once again, the university's recourse to Facebook pages for distributing educational events suggests a lack of engagement by both teachers and students with such platforms.

It could be argued that the lack of familiarity and reluctance to use e-learning platforms are also influenced by students' inability to access their login credentials. Many students express frustration with the university's inadequate technological support in addressing such issues. One participant lamented, "... knowing they put the lessons on the platform, but putting the user's name and password as mandatory, which we do not have because the administration did not open accounts for us despite their promises". Survey data further corroborates this sentiment, with over 51% of participants indicating that the university's response to technical breakdowns is weak, and more than 15% describing it as very weak. This shows the widespread dissatisfaction among students regarding the university's handling of technical issues, which in turn contributes to their reluctance to fully engage with e-learning platforms.

4.12.4 "I Believe There Is a Room for Improvement"

This theme reveals a prevailing inclination among participants towards addressing the challenges confronting e-learning platforms and distance education. These proposed solutions are rooted in the participants' experiences with diverse e-learning scenarios within their respective institutional contexts, thereby reflecting an institution-centric perspective. In articulating their suggestions, participants emphasise the imperative of mitigating the proliferation of students to enhance the efficacy of e-learning experiences. The substantial surge in student enrolment, exceeding 1.6 million students in the preceding academic year (MESRS, 2022), underscores a pressing concern vis-à-vis the inadequacy of available resources and facilities, encompassing laboratories, internet accessibility, financial support, scholarships, and opportunities for training abroad. Despite the acknowledged economic ramifications associated with accommodating a large number of graduates in the job market, as expounded upon in Chap. 2, the deleterious impact on learning resources and teaching quality prior to graduation assumes paramount significance. Furthermore, participants advocate for a strategic reassessment of e-learning modules, citing instances where module redundancy has compromised the educational experience. One participant asserts, "We have three e-learning modules and no one told us until now, we did not study anything all the semester", highlighting a disconnect between module offerings and pedagogical efficacy. Another participant underscores the nuanced nature of subject matter suitability for online delivery, stating, "I think not all subject matters (modules) can be taught electronically". This sentiment is echoed by another participant who asserts that "not all modules can be taught online", revealing the need for discernment in module selection to optimise the effectiveness of e-learning initiatives.

Additional perspectives from participants, particularly educators, further unveil the challenges posed by burgeoning student numbers, prompting calls for a reduction in teaching loads to better accommodate the administrative demands and student needs inherent in e-learning environments. These educators posit that a decrease in workload would afford them the requisite time and bandwidth to curate high-quality learning materials and deliver effective online instruction. One teacher advocates for this approach, stating, "Reduce the workload for teachers so that they can prepare and deliver online lessons. Otherwise, the experience will not be that successful", stressing the pivotal role of teacher preparation in the success of e-learning initiatives. Echoing this sentiment, another educator contends, "Reduce workload for teachers…they…can allocate time and efforts to make it (i.e. e-learning) a successful project", highlighting the potential for workload reduction to enhance the efficacy of e-learning endeavours. In the Algerian context, university teachers typically carry a teaching workload of nine hours per week, equivalent to two to three days, exclusive of the considerable time invested in lesson preparation. Moreover, teachers often find themselves shouldering additional administrative responsibilities, such as exam supervision, grading, and student record management, further exacerbating their workload burdens. These administrative tasks, originally the purview of administrative staff, have been delegated to teachers due to resource constraints and organisational pressures. One teacher argues that it is the job of a secretary's office to enter students' marks into the PROGRES platform complaining about additional duties assigned to them, "…enter the marks in PROGRES (which is the task of a secretary)…". Participants contend that reallocating administrative duties away from teachers would alleviate their workload strain, enabling them to focus squarely on instructional activities and thereby fostering improvements in distance learning practices.

Beyond local e-learning dynamics, participants critically examine the overarching dependence of universities on government funding and regulatory frameworks, positing that greater institutional autonomy could pave the way for future enhancements in HE. One participant advocates for university autonomy from government sponsorship, asserting, "Make the university independent of the government sponsoring so that it can promote its own plans on learning", highlighting the potential for self-directed institutional strategies to foster educational innovation. Resonating with this claim, another participant stresses the multifaceted interplay between e-learning and broader institutional dynamics, stating, "It is not only about e-learning but e-learning is linked to other things as well like administration and central government", emphasising the interconnectedness of e-learning initiatives with institutional governance structures. Participants assert that autonomy in decision-making represents a pivotal solution for enhancing the educational landscape in HE. As previously discussed in this chapter, the Algerian educational framework is characterised by stringent regulatory standards, constraining institutional flexibility in resource management and curriculum development. Participants lament the lack of autonomy in decision-making, citing instances where curricular changes are implemented without adequate consultation, as articulated by one teacher, "Some modules went from in-person 1.30-hour timing sessions to 3 hours without giving any explanation!" While autonomy represents a long-term solution, it is unlikely to attain the feasibility of effecting meaningful change within the current institutional framework due to the complex political realities of the country.

4.13 Conclusion

In this chapter, we looked into the digitalisation intentions of the Ministry of Higher Education in Algeria. While the government's aim to digitalise HE was commendable, the approach appeared to be overly centralised and inflexible, lacking the necessary adaptability for local institutions to address their unique challenges. The efforts primarily focused on digitalising administrative processes, neglecting crucial mechanisms for e-learning implementation. Given the significant student population across various Algerian institutions, challenges in effectively implementing e-learning initiatives became evident nationwide. The inadequacy of technological infrastructure emerged as a primary obstacle, contributing to the failure of e-learning initiatives. These shortcomings directly impact the long-term objectives of SD in HE. One of the challenges would include students graduating without proper utilisation of computers or electronic devices. This shortfall extends to the job market and has tremendous effects on the economy as well. As highlighted in Chap. 2, the United Nations established SD goals to promote the well-being, equity, and environmental responsibility of future generations. However, educational equity remains a critical criterion that many public universities in Algeria struggle to achieve. Disparities in students' family incomes underscore the unequal learning opportunities available, further exacerbating the issue. The lack of educational equity undermines education for sustainable development (ESD) in Algeria at various levels. Addressing these disparities is crucial for ensuring that all students have equal access to quality education, thereby fostering SD and societal progress.

Furthermore, due to the predominant focus on administrative aspects, digitalisation in HE often neglected the development of digital learning platforms. Consequently, universities served primarily as governmental instruments to showcase Algeria's purported strides towards SD in HE. However, a closer examination reveals that only a limited number of studies have been made in integrating SD principles within HEIs in the Arab context, including countries like Algeria as we have discussed in Chap. 2, Sect. 2.6. The absence of comprehensive frameworks and initiatives impedes the integration of SD practices within Algerian HEIs, hindering progress towards achieving SD goals in the education sector. Moreover, cultural and institutional barriers within the Algerian educational landscape may also contribute to the limited adoption of SD principles. Traditional teaching methods and curriculum structures may resist change, making it challenging to incorporate innovative approaches that promote sustainability. Additionally, insufficient investment in faculty development and training programmes focused on ESD further exacerbates the issue, leaving educators ill-equipped to integrate SD concepts into their teaching practices.

On top of that, the analysis conducted in this chapter, leveraging data from surveys, highlights the Ministry of Higher Education's acute awareness of the challenges confronting HEIs while simultaneously contributing to their decline. This revelation becomes especially pronounced in the fifth chapter, particularly in conjunction with the policy of English as a medium of instruction (EMI). The Ministry's stance of safeguarding students and affording them special privileges, as explored in Sect. 4.12.1, prompts inquiries into its underlying motives. Dris-Ait Hamadouche et al. (2022) posit that the university has metamorphosed into a political instrument, shedding light on possible motivations behind such actions. Nevertheless, numerous unanswered questions persist regarding this disconcerting state of affairs—queries that few are willing to confront.

References

Algerie Telecom. (2024). *Message from the president and CEO*. https://www.algerie-telecom.dz/en/page/mot-du-president-directeur-general-p141

An, Y. J., & Reigeluth, C. (2011). Creating technology-enhanced, learner-centered classrooms. *Journal of Digital Learning in Teacher Education, 28*(2), 54–62.

Didham, R. (2018). Education for sustainable development and the SDGs: Learning to act, learning to achieve. *Policy brief; Advancing ESD policy, UNESCO global action program on education for sustainable development*. https://en.unesco.org/sites/default/files/gap_pn1_-_esd_and_the_sdgs_policy_brief_4_page_version.pdf

Dris-Ait Hamadouche, L., Oussedik, F., & Taleb Ibrahimi, K. (Eds.). (2022). *L'université désacralisée: Recul de l'éthique et explosion de la violence*. Koukou.

Ezeofor, I., & Lent, R. W. (2014). Social cognitive and self-construal predictors of well-being among African college students in the US. *Journal of Vocational Behavior, 85*(3), 413–421.

Fatahi, S., & Moradi, H. (2016). A fuzzy cognitive map model to calculate a user's desirability based on personality in e-learning environments. *Computers in Human Behavior, 63*, 272–281.

Foucault, M. (1991). Governmentality. In G. Burchell, C. Gordon, & P. Miller (Eds.), *The Foucault effect: Studies in governmentality* (pp. 87–104). The University of Chicago Press.

Goedegebuure, L., & Hayden, M. (2007). Overview: Governance in higher education – Concepts and issues. *Higher Education Research & Development, 26*(1), 1–11.

Grimes, N. D., & Porter, W. (2023). Closing the digital divide through digital equity: The role of libraries and librarians. *Public Library Quarterly*. https://doi.org/10.1080/01616846.2023.2251348

Harmon, J. C. (2007). Let them use the internet: Why college instructors should encourage student internet use. *College Teaching, 55*(1), 2–4.

MESRS. (2022). *Grands agrégats* (Large aggregates). https://www.mesrs.dz/index.php/fr/agregats-2/

Palumbo, E., & Verga, F. (2015). Creation of an integrated environment to supply e-learning platforms with Office Automation features. *Interactive Learning Environments, 23*(6), 766–777.

Sayaf, A. M. (2023). Adoption of e-learning systems: An integration of ISSM and constructivism theories in higher education. *Heliyon, 9*(2), 1–17.

Schmidt-Crawford, D. A., Lindstrom, D. L., & Thompson, A. D. (2023). The ISTE pledge: Committing to digital equity and transformation. *Journal of Digital Learning in Teacher Education, 39*(1), 2–3.

Trend, R., Davis, N., & Loveless, A. (1999). *QTS: Information and communications technology*. Letts.

UNESCO. (2018). *Paving the road to education: A target-by-target Analysis of SDG 4 for Asia and the Pacific*. https://unesdoc.unesco.org/ark:/48223/pf0000265912

Wilson, C. (2013). *Credible checklists and quality questionnaires: A user-centered design method*. Elsevier.

Zalazar-Jaime, M. F., Moretti, L. S., García-Batista, Z. E., & Medrano, L. A. (2023). Evaluation of an academic satisfaction model in e-learning education contexts. *Interactive Learning Environments, 31*(7), 4687–4697.

Zhou, Y., Zhao, J., & Zhang, J. (2020). Prediction of learners' dropout in e-learning based on the unusual behaviors. *Interactive Learning Environments, 31*(3), 1796–1820.

CHAPTER 5

English as a Medium of Instruction in Practice: Exploring Implementation and Impact

5.1 Introduction

The desire to introduce English as a medium of instruction (EMI) in Algerian higher education (HE) has been a continuous pursuit since 2019, as highlighted in our previous discussion in Chap. 3. The primary objective behind this initiative is to enhance the international visibility and recognition of Algerian universities on the global stage. A notable strategy employed to facilitate this transition is the promotion of English language in primary schools, serving as a foundational step to prepare students for EMI at HE levels. However, the effectiveness of implementing EMI in Algerian HE is contingent upon various factors, notably the government's approach to this initiative. Critical components include the formulation of comprehensive strategies, provision of adequate facilities, and development of supportive policies tailored to facilitate the successful integration of EMI. In this regard, the government's strategies encompass not only the promotion of English language proficiency among students and educators but also the establishment of infrastructure and resources conducive to EMI delivery. This entails investment in language learning programmes, teacher training initiatives, and curriculum development efforts aimed at aligning educational objectives with good standards. Furthermore, supportive facilities such as language laboratories are essential to facilitate interactive and immersive English language. Additionally, policies addressing curriculum adaptation and assessment methodologies play a pivotal role in ensuring the quality and sustainability of EMI implementation.

Following the decision to explore EMI in HE, the establishment of local and national committees marked a significant step towards overseeing the implementation of this initiative. These committees, composed of stakeholders from academia, played a crucial role in providing guidance, coordination, and support for the integration of EMI. The formation of these committees lent

S. Bouherar, S. Salem, *Sustainability in Algerian Higher Education*,
https://doi.org/10.1007/978-3-031-81347-4_5

credibility to the decision to adopt EMI and emphasised the seriousness of the endeavour. By bringing together experts and representatives from various institutions, these committees contributed to the development of comprehensive strategies, frameworks, and action plans for the effective implementation of EMI across Algerian universities. However, it is important to acknowledge that while these committees may have strengthened the decision and provided valuable input, they may have limitations in terms of authority to enforce compliance with ministerial instructions. While they can offer recommendations, guidance, and support to institutions, they may lack direct disciplinary powers to sanction non-compliant institutions, as their job is to report the progress to higher ministerial authorities. Institutions may vary in their capacity and willingness to comply with mandated changes, and without robust mechanisms for accountability and enforcement, the efficacy of these committees in driving widespread adoption of EMI may be limited.

The government's efforts to promote and implement EMI in the HE sector in Algeria are commendable. However, the effectiveness of these efforts may be hindered by organisational and conceptual challenges that impede the proper implementation of this decision. One significant issue is the lack of accurate statistics regarding the implementation of EMI, including the availability of programmes, the progress of teachers, and the proficiency levels of students in English, as well as their readiness to embrace this change. The absence of comprehensive data on EMI implementation makes it difficult to assess the extent to which the initiative is being successfully integrated into higher education institutions (HEIs). Without accurate statistics, it is challenging to identify areas of improvement, measure progress, and allocate resources effectively. Moreover, the focus of training programmes primarily on educators rather than students may overlook the importance of students' readiness and proficiency in English. While equipping educators with proficient English language skills is essential for effective instruction, the success of EMI ultimately depends on the ability of students to understand and engage with the material presented in English.

The conceptualisation of EMI by the Ministry of Higher Education in Algeria, particularly in relation to English for Specific Purposes (ESP), is crucial for the effective integration of EMI into teaching practices. ESP focuses on developing English language skills specific to particular academic disciplines or professional fields, ensuring that educators are equipped with the language proficiency and pedagogical strategies necessary to teach their subjects effectively in English. In contrast, EMI encompasses a broader approach to using English across various academic subjects, without explicitly targeting the specific language needs of individual disciplines. While ESP emphasises the adaptation of language instruction to suit the linguistic and academic requirements of specific fields, EMI focuses on the overall delivery of academic content in English, regardless of the subject matter. Another related concept is Content and Language Integrated Learning (CLIL), which aims to develop both language proficiency and subject knowledge simultaneously, offering students opportunities to learn academic content through the medium of a foreign language, such as English. In other words, does the Ministry provide programmes tailored to educators' subject-based needs to effectively integrate EMI into their teaching practices? How is EMI similar or different from ESP and CLIL as integrated in Algerian tertiary education? In response to these challenges, this chapter discusses shortages surrounding the implementation of EMI and addresses the Algerian Ministry of Higher Education view of EMI through the current textbooks and curriculum used to train teachers. The discussion this chapter provides prompts us to reconsider the necessity of a multilingual approach to education for Algerian students, particularly in light of concerns that the current use of EMI may inadvertently lead to exclusion rather than inclusion, which would contradict the targets of Sustainable Development Goal 4 (SDG4) (see Sect. 2.3).

5.2 Multilingualism Intentions, Monolingualism Practices

Whether the Ministry has opted for ESP or EMI with or without a framework of CLIL, it appears that the government's intentions lean towards embracing a multilingual approach in education, with its previous history of using French and Arabic as a medium of instruction as we have illustrated in Chap. 3. This inclination is observable in the bilingual approach adopted in higher education, particularly in the teaching of hard sciences versus social sciences, as highlighted by Bouherar and Ghafsi (2021) and Alkhateeb and Bouherar (2022). Additionally, socio-linguistic dynamics in Algeria seem to align with a multilingual ethos (Belmihoub, 2018), with colloquial Arabic, Tamazight, and French serving as primary languages that define Algerian societies (Bouherar & Ghafsi, 2021). However, the multilingualism evident in Algerian societies and attempted in the HE system through the use of EMI may differ from the concept of multilingualism characterised by socio-political dynamics such as "minority" and "majority", "us" and "them", and "standard" and "non-standard varieties", as discussed by Weber and Horner (2012). These dichotomies are already inherent in the socio-linguistic landscape of Algeria, as elucidated by scholars like Benrabah (2005), Tilmatine (2015), and Le Roux (2017). The question arises: has the government acknowledged and addressed the complexity and problematic nature of concepts like race, ethnicity, minority, and language ideologies surrounding the use of multilingualism in Algerian higher education? This section aims to uncover and discuss the government's approach to these issues.

The intricate glotto-political landscape of Algerian society (Belmihoub, 2018) has remained in flux since independence, stemming from the enduring effects of colonial policies (Bouherar & Ghafsi, 2021). This ongoing political strife has permeated all facets of social existence, particularly education, where language choices are predominantly dictated by government mandates with little grassroots input (Benrabah, 2005). The government's unilateral stance on linguistic matters has started tensions among three main linguistic groups: Arabophones, Berberists, and Francophones, with the relationship between Arabophones and Francophones historically marked by friction. Until the mid-1970s, French coopérants largely controlled Algeria's education system (Kadri, 2014), shaping it according to their own perceptions of the country's future trajectory (see Chap. 3). The brewing conflict between Arabic and French reached a crescendo with the government's Arabisation efforts in the mid-1970s, spurred on by demands from Islamic schools (Bennoune, 2000). This shift from French to Arabic as the medium of instruction prompted the departure of French coopérants, signalling the perceived demise of their linguistic and cultural influence, although these tensions persist. The clash of ideologies inherent in this linguistic conflict is exemplified by the experiences of prominent national figures like former Ministers of National Education, Nouria Benghebrit and Ali Ben Mohamed, whose contentious interactions underscore the enduring divide between Arabophones and Francophones, as evidenced by Benghebrit's decision to bar Ben Mohamed from delivering a talk on Arabic language education during colonisation in Algeria back in 2014.

The Tamazight language, primarily spoken by Berbers, has been embroiled in significant turmoil as Berbers strive to assert the recognition of their language and culture. Events such as the Berber Spring have showcased Berberists' resistance against government suppression of Tamazight culture, while movements like the Arouch Movements in 2001 protested the government's policies toward the Kabyle region. The tragic death of the renowned Kabyle singer Matoub Lounes has also sparked violent outbursts, reflecting deep-seated tensions and expressions of resentment against the dominance of the Arabic language (Benrabah, 2005). Within this context, multilingualism is framed as a battleground between minority and majority, oppressor and oppressed, dominant and submissive, with Berberists casting themselves as victims and perceiving the Arabised government as the oppressor.

While Algeria officially recognises only two languages, Arabic and Tamazight, in its constitution, the socio-political dynamics and historical legacy have resulted in a de facto multilingual environment, with French playing a significant role in education and literature (Daoudi, 2018). Despite not being officially designated as an official language, French remains widespread and holds a high status, often surpassing the local languages in terms of usage and opportunities. Algerian elites often prioritise French-based education for their children (Benrabah, 2013), leading to preferential treatment in employment and other opportunities (Daoudi, 2018). This unique multilingual situation in Algeria diverges from examples like Switzerland, where multilingualism is fostered through a federalist approach, with each region having its primary language and others being secondary (Weber & Horner, 2012). However, in Algeria, the status of Tamazight is evolving, with efforts to expand its recognition beyond its traditional strongholds, albeit facing resistance, particularly in areas where Arabic dominance has been established. Despite the linguistic tensions, French serves as a common link between the two official languages, spoken by both ethnicities, further complicating the multilingual landscape of the country.

The complex multilingual landscape in Algeria significantly influences language use in education, with tensions arising from both Berberists challenging the imposition of Arabic in their communities and Arabs resisting the teaching of Tamazight in their schools. This contentious debate resurfaced notably when the Algerian parliament rejected the broader integration of Tamazight into the national education system in 2017, sparking division among Algerians. Despite the government's constitutional recognition of Arabic and Tamazight as official national languages, the usage of Tamazight remains predominantly regional, driven by ethnic ideologies. In the realm of education, Tamazight is offered as a major at universities in Batna, Bejaia, Tizi-Ouzou, and Bouira, while Arabic serves as the medium of instruction in social sciences and French in scientific and technological domains (Bouherar & Ghafsi, 2021). Amidst this socio-political approach to multilingualism in education, it is crucial to acknowledge the pivotal role of the French language. The language's status is underscored by its cultural, economic, and political significance, shaping opportunities and interactions for its speakers (Mackey, 1989), and that clearly defines the status of French in Algeria. Despite the intricate linguistic dynamics, the Algerian government, through the Ministry of Higher Education, has opted to introduce EMI in tertiary education, reflecting a commitment to a multilingual approach and responding to globalisation pressures, while also challenging the dominance of French (Jacob, 2020). This recent shift in language policy in HE can also be viewed as part of the broader decolonisation efforts within the education system (Bouherar & Ghafsi, 2021). However, concerns arise regarding the proficiency levels of students in English, highlighting that this decision transcends regional considerations common in other multilingual countries and holds national implications. As a consequence, the move towards incorporating English may inadvertently perpetuate the ideology of monolingualism, potentially sidelining French in favour of another linguistic dominance.

5.3 English Language Approaches in Education: ESP, EMI, and CLIL

Language functions as the principal medium of human communication, facilitating interpersonal connectivity and the dissemination of ideas. Our global landscape is characterised by linguistic diversity, encompassing a multitude of languages. Within many nation-states, a designated national language coexists alongside a variety of indigenous languages spoken by local communities. Historical colonial legacies have introduced secondary languages, while the process of globalisation has brought additional foreign languages. Globalisation involves the global movement of technology, information, goods, and people across borders, reshaping economies, societies, environments, cultures, and HE systems worldwide. This development refers to "the shrinkage of distance and time-delay in communications and travel, leading to increasingly extensive and intense global relations" (Marginson & Rhoades, 2002, p. 288). There is an ongoing debate surrounding the definitions of globalisation and internationalisation. Marginson (1999) distinguishes internationalisation as involving bilateral or multilateral relationships between individual nations, with the nation-state serving as the fundamental unit. Globalisation, on the other hand, encompasses "world systems" that often influence local or national dynamics. This does not imply an absolute or one-sided dominance of global forces over national and local ones; rather, it suggests the potential for global factors to impact all parts of the world. Marginson illustrates this with the example of modernism, which is no longer solely associated with individual nation-states.

Marginson and Rhoades (2002) introduce a heuristic framework termed "glonacal agency" within the field of comparative HE research. Their primary aim is to emphasise the significance attributed to the examination of global phenomena. However, they refrain from regarding such phenomena as universally applicable or deterministic in their effects. Instead, they highlight the enduring relevance of the national dimension. They eschew the portrayal of global or national phenomena as all-encompassing in their impacts, thereby stressing the importance of the local dimension. They coined the term "glonacal" to encapsulate the intricate interplay among global, national, and local dynamics. Applying this transformative perspective in the context of the English language, globalisation affects educational institutions by encouraging the teaching of English as a second language or even as a primary medium of instruction. This shift does not mean that national and local languages are being eradicated, but rather that English is becoming an additional tool for communication, learning, and market access in an increasingly interconnected world. Gopinathan (2007, p. 55) asserts:

> Globalization's effect can be seen in greater internationalisation, of a trend towards the commodification of education, greater convergence in views about how education should contribute to the economy, greater use of choice, competition, deregulation and increasing both the involvement and burdens for parents.

Globalisation has profoundly influenced education by promoting increased internationalisation, facilitating the mobility of students, educators, and educational resources across borders, thereby enriching academic institutions with diverse cultures and perspectives. However, it has also resulted in the commercialisation of education, accentuating elements such as branding, marketing, and profit-making within the educational sector. This trend underscores the crucial role of education in fostering economic growth and development, leading to an emphasis on skill enhancement and workforce training. For instance, in 2020, Algeria implemented an innovative economic model that included the establishment of university business incubators to attract students towards entrepreneurship. As part of their graduation projects, university students are encouraged to establish start-ups and micro-enterprises, utilising state-provided funding resources to contribute significantly to economic development initiatives. There has been a shift towards market-oriented policies in education, which has increased the responsibilities and involvement of parents in their children's educational journey. Globalisation has driven the internationalisation of education, introducing policies and practices influenced by the global market. Educational institutions are rapidly embracing internationalisation, particularly through increased utilisation of new communication technologies and the integration of EMI (O'Neill & Chapman, 2015). Algeria's recent educational reforms, specifically the implementation of digitalisation and the introduction of EMI, can be viewed as a strategic move towards internationalising its HE system.

In 2019, the former Minister of Higher Education in Algeria Tayeb Bouzid sought to draw public attention towards the imperative need for replacing French with English as the primary language of research and instruction in universities. This initiative aimed to enhance visibility and foster international engagement, recognising English as the global lingua franca. The overwhelming majority of Algerians positively responded to this proposition, signalling a willingness to adopt English as their primary foreign language over French. Subsequently, training programmes were initiated for educators, encouraging them to utilise EMI in their teaching. In furtherance of the internationalisation of higher education, the Ministry introduced the "Study in Algeria, One Star" label in 2022 to attract foreign students. At the 2023 ceremony, this label was conferred upon 15 university establishments that met the requisite educational and service standards for accommodating international students. Valid for one year, this label serves to incentivise universities to sustain their efforts in enhancing the quality of services offered. As part of a broader digitalisation policy extending to all state sectors, the Ministry advocated for the integration of modern technologies within universities to strengthen their internationalisation efforts. In 2022, Minister Kamel Baddari announced the "zero paper" policy, requiring that all university theses, research, and documentation be submitted via modern technological platforms. At the beginning of the academic year 2023–2024, this policy was commemorated as students completed their registrations online, notably utilising platforms such as PROGRESS. The digitalisation drive within HE not only augmented its appeal to foreign students but also facilitated the ascendancy of EMI in Algeria. Beyond the objectives of elevating the visibility of Algerian research and education, the transition to English, which commenced during the Al Hirak movement, can be perceived as a means of decolonising education and diminishing reliance on the colonial language, French.

However, it cannot be overlooked that English has emerged as one of the most widely spoken languages globally, largely due to historical colonial influences. The expansion of the British Empire into numerous territories resulted in the imposition of English upon indigenous populations. Many of these countries have retained English as a secondary language, with some even adopting it as their official language. Today, English speakers can be found in almost every corner of the world, establishing it as the predominant lingua franca, extending beyond areas directly influenced by British colonialism. It serves as the primary language for commerce, tourism, technology, and scientific discourse. Numerous nations have integrated English into their educational systems, recognising its vital role in fostering socio-economic development and expanding employment opportunities. The landscape of English language education has undergone significant transformation on a global scale. Initially focused on teaching English as a second language (ESL) and foreign language (EFL), it has since evolved to encompass English for Specific Purposes (ESP), followed by Content and Language Integrated Learning (CLIL), and ultimately the adoption of EMI. ESP, CLIL, and EMI are considered to be grounded in Content-based Instruction (CBI) (Carrió-Pastor, 2021). These approaches signify substantial progressions in language education. CBI, an American Canadian concept, is an overarching approach that involves "the integration of language teaching aims with subject matter instruction" (Snow, 2001, p. 303). The incorporation of English into the educational system in Algeria can be observed as moving from EFL to including ESP in EFL classes, then the implementation of EMI in content classes, and discussions among researchers about CLIL and its potential to enhance both language proficiency and subject content within EMI classes.

With English emerging as the global lingua franca for fields such as science, technology, and business, particularly ignited by the oil crises of the early 1970s which prompted a substantial influx of funds and Western expertise into oil-rich nations, a distinct category of learners appeared with their unique motivations for learning English. This growing demand led to the development of a specialised approach to English language instruction tailored to the specific needs of learners, known as ESP. The guiding principle of ESP became encapsulated in the phrase, "Tell me what you need English for, and I will tell you the English that you need" (Hutchinson & Waters, 1987, p. 8). As it became evident that there were significant differences in the English language usage across various academic and occupational fields, specialised branches of ESP began to arise, such as English for Science and Technology (EST) and English for Business and Economics (EBE). What sets ESP apart from general English courses is the underlying assumption that ESP students already possess a certain level of proficiency in English, and the primary goal of language learning is effective communication within a specific domain. ESP places emphasis on the language, vocabulary, grammar, discourse styles, and genres specific to particular disciplines, rather than relying on general grammar rules, learner's dictionaries, and everyday discourse (Räisänen & Fortanet-Gómez, 2008). The provision of ESP courses at the university level has witnessed significant growth over the years. Räisänen and Fortanet-Gómez (2008, p. 12) assert that "most of the English taught at universities in Europe today is English for Specific Purposes (ESP)", highlighting the prevalence and importance of this approach.

In Algeria, ESP is integrated into EFL classes, as students are expected to enter university with a foundational knowledge of English from their prior education.

Recently, there has been a rapid global transition in non-Anglophone nations from teaching EFL to adopting EMI for academic subjects like science, mathematics, geography, and medicine. EMI, defined as "the use of the English language to teach academic subjects in countries or jurisdictions where the first language (L1) of the majority of the population is not English" (Dearden, 2014, p. 4), is gaining prominence worldwide. While the surface reasons for the growth of EMI may seem similar across contexts, deeper examination reveals a myriad of contextual, geographical, historical, and political factors shaping each country's adoption of EMI (Dearden, 2014). For instance, in Algeria, a multilingual nation, the impetus behind introducing EMI appears to be contextual and political. Contextually, it aims to enhance visibility, attract international students, and improve students' career prospects and national development. Politically, it seeks to reduce the dominance of the colonial language, French. Beyond the internationalisation of HE and the enhancement of career opportunities, EMI's prevalence is also attributed to its dual benefits of acquiring English proficiency and subject knowledge simultaneously (Rose et al., 2019). To fully leverage the potential of EMI, stakeholders need to develop strategies to promote language awareness and ensure teaching quality. However, EMI primarily focuses on subject matter instruction in English without explicit language learning objectives (Dafouz, 2018). Unlike ESP, which emphasises language learning, EMI prioritises disciplinary-content learning, taught and assessed by subject experts (Dafouz, 2021). Despite these distinctions, ESP and EMI share a common emphasis on integrating content and language, albeit to varying degrees. Costa and Mastellotto (2022) advocate for the integration of ESP and EMI, considering them complementary. Therefore, collaboration between ESP and subject teachers is crucial in Algeria to ensure the quality of teaching and learning.

Besides discussing EMI and ESP, there is also a debate about the extent to which these concepts differ from closely related ones such as CLIL. CLIL, generally defined as "a dual-focused educational approach in which an additional language is used for the learning and teaching of both content and language" (Coyle et al., 2010, p. 1), gained popularity worldwide due to globalisation and internationalisation. It became widely adopted in secondary and primary schools as well as universities, practised in both content and language subjects. While CLIL enhances English language learning and yields positive outcomes in content classes, it can also pose challenges, potentially diverting attention from core module objectives and exacerbating English language deficiencies, particularly for academically, linguistically, or economically disadvantaged students (Bruton, 2011, 2013). In contrast, Hüttner and Smit (2014) argue that CLIL enhances students' lexical knowledge in specific subjects, boosts engagement and content learning, and increases motivation to learn English, aligning with the necessity of English proficiency in our globalised world. CLIL garnered attention alongside EMI. While CLIL emphasises teaching and learning both content and language without specifying the language of instruction, EMI stipulates English as the language of education, though not necessarily prioritising it as a subject. CLIL primarily found application in primary and secondary schools, whereas EMI initially emerged at the university level. While CLIL is contextually situated, rooted in the European concept of plurilingual competence for EU citizens, EMI lacks specific contextual origins (Dearden, 2014; Carrió-Pastor, 2021).

Differentiating between CLIL and EMI, Carrió-Pastor (2021) asserts that both are forms of bilingual education. However, while CLIL involves teaching content through any foreign language, EMI entails teaching content to students proficient in English. In CLIL, teachers have a dual objective of teaching both language and subject content, whereas in EMI, content teachers primarily focus on delivering subject matter in a foreign language. CLIL involves assessment of both content and language acquisition, whereas EMI primarily assesses content acquisition alone. Challenges arise when students lack proficiency, impacting both language and content comprehension. Consequently, EMI students with inadequate English proficiency may perceive a lack of competence in content, demotivating them and hindering both English and content acquisition. Thus, the choice between CLIL and EMI should consider students' foreign language proficiency. If students lack proficiency, CLIL can reinforce language acquisition strategies, preventing demotivation. Conversely, EMI can be offered if students demonstrate adequate language proficiency and institutions have trained teachers. Successful implementation depends on student enrolment, teacher skills, and careful consideration of language proficiency. Ignoring these aspects may lead to students' inadequate proficiency in both English and content, discouraging foreign language use. Airey's (2016) continuum of approaches in HE places CLIL between the language-focused ESP and the content-focused EMI. Algeria's adoption of EMI may be rooted in the assumption that students possess English proficiency from middle and secondary school, a trend recently extending to primary education. However, many students and even teachers lack adequate English proficiency, especially in scientific branches where English was secondary. Teacher training solely focused on enhancing English proficiency may not adequately prepare them for successful EMI implementation, particularly after prolonged use of French. Integrating CLIL into EMI classes could enhance both content and language learning, warranting further research.

Throughout this book, we employ the concept of EMI to reflect recent reforms in multilingual Algeria. Multilingual learners often utilise their diverse language repertoires to facilitate learning. They select from their verbal repertoires, which include "all the accepted ways of formulating messages" according to the meaning they wish to convey (Gumperz, 1964, p. 138). This ability, recently termed "translanguaging", refers to "the ability of multilingual speakers to shuttle between languages, treating the diverse languages that form their repertoire as an integrated system" (Canagarajah, 2011, p. 401). Translanguaging supports multilingual learners in language and content acquisition, as well as in developing multilingual awareness (Smit, 2019). In EMI classes, translanguaging, combining English with other languages shared by some or all multilingual actors, fulfils various functions, supporting discourse processes, meaning-making, content comprehension, educational quality, classroom interaction, and enhancing employment opportunities (Van der Walt, 2013; Smit, 2019). Despite English being assumed in EMI, research into classroom practices indicates that lecturers and students engage in translanguaging between English and other languages. Branka and Molino (2022) found that lecturers from five European universities use English for its contribution to an international experience and for students' language skill development. Additionally, they use other languages for compensation, pedagogical purposes, accommodation, and identity expression.

While multilingual practices are often overlooked, many teachers and students in EMI classes are likely actively cultivating flexible, multilingual environments rather than confining themselves solely to English. This trend is perhaps especially notable in multilingual Algeria, where EMI is being implemented in classrooms that were traditionally taught predominantly in French, despite Arabic and Tamazight being recognised as official languages of the country. Algeria's multilingual context suggests that adopting an English-exclusive stance could potentially marginalise indigenous languages and exclude students with inadequate English proficiency, contradicting SDG4's emphasis on inclusiveness and quality education. Branka and Molino (2022) advocate for the emergence of multilingual-oriented EMI policies from classroom communication needs, cautioning against top-down imposition. Le Hoang Ngo (2021) similarly discovered that students in an EMI programme at a Vietnamese University employ their linguistic repertoires to create meaning and develop content knowledge, arguing that these translanguaging practices contribute to the formation of bottom-up policies at the classroom level, which may or may not align with top-down policies. Refraining from allowing the use of languages other than English restricts access to knowledge, undermines social justice, and hinders effective learning (Van der Walt, 2013). Adopting a multilingual approach that utilises translanguaging to meet various classroom needs in EMI classes in Algeria has the potential to enhance both language and content learning, thereby fostering the sustainability of the EMI reform. This chapter examines EMI implementation in Algeria through the lens of sustainability, considering perspectives of teachers and students.

5.4 Research Design

The research design serves as a methodological framework structured to systematically address the study's inquiries (Yin, 2003). Its primary aim is to attain a thorough and nuanced understanding of the phenomenon under investigation. In this study, an exploratory case study design is chosen to delve into the central research question: what are the perspectives of Algerian university teachers and students regarding the integration of EMI within Algerian HEIs?

The selection of an exploratory case study design is predicated on its ability to investigate complex social phenomena within their authentic contexts

(Stake, 1978). This approach enables a comprehensive exploration of the viewpoints, experiences, challenges, and potential benefits associated with EMI, as articulated by both teachers and students. By focusing specifically on the unique landscape of Algerian higher education, this research aims to uncover insights that may not be fully captured by other research methodologies. This design facilitates a deep exploration that goes beyond surface-level understanding, aiming to illuminate the intricacies of how EMI is viewed within Algerian universities and offering valuable insights into its practical implications and potential outcomes upon integration.

Given the research's focus on exploring participants' perspectives on EMI integration in Algerian universities, a qualitative approach is employed due to its emphasis on understanding how individuals interpret their world (Bryman, 2016). This approach focuses on the collection and analysis of textual data, allowing for a detailed examination of participants' voices. Through qualitative exploration and contextual analysis, the study seeks to provide detailed perspectives that can inform educational policies and practices in Algerian HEIs. This approach not only captures the richness of personal experiences and views but also contributes to a broader understanding of the factors influencing EMI adoption in Algeria.

This study's design and approach are rigorously selected and implemented to offer a deep and insightful exploration of the complexities surrounding EMI integration in Algerian higher education. It aims to provide valuable empirical evidence that can assist stakeholders in balancing educational reforms, practices, and outcomes within Algerian HEIs.

5.4.1 Participants

Algerian university educators and students from various disciplines were intentionally invited to share their perspectives on the recent initiative to introduce EMI in Algerian HEIs. Purposive sampling, which involves selecting participants based on specific criteria relevant to the research (Denscombe, 2007; Etikan et al., 2016), was employed to effectively meet the study's objectives. Participants were chosen irrespective of their university affiliation or academic field to ensure a comprehensive and diverse range of viewpoints on the subject under investigation.

Convenience sampling, another strategy used to select participants based on their availability and willingness to participate, was also utilised to gather rich data. Although a substantial number of Algerian university teachers and students were invited to partake in the study, only 23 teachers and 8 students voluntarily agreed to participate. The participants were predominantly from the eastern, western, and northern regions of Algeria, as illustrated in Table 5.1.

The geographical locations of the participants are not a critical factor in this study; however, our aim was to incorporate perspectives from teachers and students across varied locations to offer a comprehensive portrayal of the integration of EMI in Algerian higher education.

Table 5.1 The locations of participants

Location	*Number of teachers*	*Number of students*
North	5	0
East	14	5
West	4	3
Total	23	8

5.4.2 *Data Collection*

To achieve this study's objectives effectively, it is crucial to gather information accurately from pertinent sources, ensuring the trustworthiness of outcomes and facilitating informed conclusions and recommendations. This research adopted a qualitative approach to delve deeply into participants' perspectives on the incorporation of EMI in Algerian HEIs, a topic underexplored, particularly concerning sustainability. Structured interviews were conducted with teachers and students to elicit detailed insights in their own words. This method involved preparing an interview schedule with predetermined questions presented in a consistent order and wording, ensuring uniformity across participants and data sources (teachers and students) (Kumar, 2011). Given the geographical dispersion of participants across different cities, online structured interviews were chosen as the mode of data collection. The recruitment process involved sharing the interview details via email and academic Facebook pages, with 31 participants ultimately agreeing to take part in the study.

Qualitative inquiries, by their nature, are subjective and context-dependent. Guba (1981) proposes four criteria for assessing their trustworthiness. Credibility, ensuring that research findings faithfully represent participants' perspectives, was achieved through triangulation. This technique involves employing multiple investigators, data sources, and methods (Bryman, 2016). This research, conducted by two researchers, utilised structured interviews with both teachers and students to cross-check its findings. Transferability, concerning the extent to which results can be applied to other contexts, was enhanced by providing detailed descriptions of the research setting and procedures, enabling readers to gauge applicability to their own situations. Dependability, focusing on the consistency of results, was ensured through reflexivity and peer examination, scrutinising the research process and findings to mitigate biases and uphold rigour. Confirmability, which safeguards that findings are influenced by participants rather than researchers' biases, was maintained by conscientiously aligning interpretations with data through reflexivity and triangulation.

This study underscores the methodological rigour essential for exploring EMI integration in Algerian HEIs. By employing structured interviews and adhering to established qualitative research criteria, the study offers robust insights into the perspectives and experiences of teachers and students. These findings not only advance understanding of challenges and opportunities in EMI implementation but also provide a foundation for future research and policy development in the evolving educational landscape.

5.4.3 *Data Analysis*

The qualitative data obtained from interviews were subjected to thematic analysis, adhering to the steps outlined by Braun and Clarke (2006). Braun and Clarke characterise thematic analysis (TA) as a flexible method for identifying and reporting themes within qualitative data. They describe themes as patterns that encapsulate significant elements of the data in relation to the research questions. The TA for this study was conducted as follows:

5.4.3.1 *Familiarising Ourselves with the Data*

Upon conducting the interviews with participants, we thoroughly immersed ourselves in the data, engaging in multiple readings until we became well-acquainted with its breadth and depth. As we carefully examined the entire dataset, preliminary ideas about potential patterns began to emerge, prompting us to initiate the coding process.

5.4.3.2 *Generating Initial Codes*

Having familiarised ourselves with the data, we began generating codes, labels identifying aspects of the data that appeared significant to the study. Our primary objective was to investigate participants' perspectives on the integration of EMI in Algeria concerning sustainability. This means that we aimed to understand how the participants viewed the use of EMI in Algerian HEIs and how this relates to sustainable practices and development within the educational system. However, we remained open to new codes emerging from the data, allowing for the identification of intriguing themes in subsequent phases. Table 5.2 indicates examples of data extracts and their corresponding codes.

5.4.3.3 *Searching for Themes*

After coding all data, we proceeded to analyse various codes, grouping them into themes and organising data extracts accordingly. For example, the codes decolonising education and decolonising scientific research were categorised under the sub-theme decolonising education and scientific research. This phase

Table 5.2 Data extracts with codes

Data extracts	*Coded for*
1. It was suggested in 2019 as a part of decolonising education in Algeria, another attempt at Defrenchification albeit it was not phrased this way. 2. The French language has lost its scientific luster and is no longer as strong as it was before.	1. Decolonising education 2. Decolonising scientific research

culminated in a comprehensive collection of sub-themes, with all data extracts coded in alignment with their respective themes.

5.4.3.4 *Reviewing Themes*

In this phase, all excerpts underwent thorough review to ensure their appropriate categorisation under respective sub-themes. The sub-themes themselves were also subject to review and consolidation to establish additional hierarchical layers of sub-themes. For instance, the sub-themes decolonising education and scientific research, and unlocking international opportunities and job market prospects were integrated under the overarching sub-theme EMI implementation drivers. By the conclusion of this stage, we had identified several cohesive patterns alongside their relevant data excerpts.

5.4.3.5 *Defining and Naming Themes*

After thoroughly reviewing all sub-themes, we proceeded to refine and analyse the data within them for relevance and consistency. This phase culminated in the identification of two main themes: the contemporary landscape of EMI adoption within Algerian HEIs and the future of implementing EMI in Algerian.

5.4.3.6 *Producing the Report*

When confronted with a series of overarching themes and sub-themes, we embarked upon composing the report, representing the final phase of data analysis. The analysis was articulated into a cohesive and engaging narrative, supported by ample data extracts as evidential support. Interpretations were undertaken in light of existing research literature, addressing the principal research question underpinning this study.

5.4.4 *Ethical Considerations*

Ethics were rigorously upheld throughout the conduct of this study. Ethical principles, as articulated by Diener and Crandall (1978), encompass four key elements: causing no harm to participants, obtaining informed consent, ensuring no invasion of privacy, and avoiding deception. This study adhered to these principles meticulously.

The research design and instruments were carefully planned and executed to prevent any physical, psychological, or other forms of harm to participants. The well-being of participants was a primary consideration throughout all stages of the research process, ensuring that no adverse effects occurred. Participants were provided with comprehensive information about the nature and purpose of the study. They received explanations regarding the research aims, procedures, potential risks, and benefits. Informed consent was obtained voluntarily, with participants indicating their agreement by ticking the informed consent box. This process guaranteed that participation was based on a clear understanding of the study and was entirely voluntary. Privacy protection for participants was rigorously maintained. Measures were taken to ensure that identities and personal information remained confidential. Anonymity was guaranteed by using pseudonyms, preventing identification. Additionally, the universities where participants worked or studied were not disclosed in the study, further safeguarding their privacy. There was no deception involved in the study. Participants were transparently informed about the study's objectives and their roles within it. Clear communication about the study's scope and intent was provided, and no misleading information was given. This transparency helped build trust and ensured that participants were fully aware of the research context. By adhering to these ethical principles, the study maintained a high standard of ethical integrity, ensuring the protection and respect of all participants involved.

5.5 The Contemporary Landscape of EMI Adoption Within Algerian HEIs

In this section, we delve into the current state of integrating EMI within Algerian HEIs. This exploration serves as a fundamental basis for depicting the EMI landscape in Algeria. As globalisation continues to shape the interconnectedness of nations, the role of English as a lingua franca cannot be understated. Its prominence extends beyond mere communication to encompass academic discourse, professional opportunities, and cultural exchange. Thus, the integration of EMI within Algerian HEIs represents a significant step towards aligning with global trends and enhancing the country's competitiveness on the international stage. Concurrently, the subsequent section explores the future trajectory of EMI within Algeria. By looking ahead, we aim to anticipate the evolving needs of Algerian students, educators, and society at large. This forward-thinking approach enables us to identify potential challenges, opportunities, and areas for growth, thereby informing strategic decision-making and policy formulation. These overarching themes, comprising main and sub-themes, will be meticulously analysed through the lens of sustainability.

Sustainability, in this context, transcends traditional notions of environmental conservation to encompass social, economic, and cultural dimensions. It embodies the imperative of meeting present needs while safeguarding the capacity of future generations to meet their own needs. As such, our analysis of EMI integration in Algerian HEIs is not limited to its immediate impact but extends to its long-term implications for educational equity, social cohesion, and economic development. This study aims to analyse the perspectives of Algerian university teachers and students within the framework of SDG4, which emphasises the necessity of inclusive and equitable education, as well as the promotion of lifelong learning opportunities for all. Through this lens, we endeavour to ascertain whether EMI practices in Algeria are fostering a sustainable and equitable educational landscape. Our inquiry extends beyond mere academic proficiency to encompass broader considerations of access, inclusion, and social justice. By interrogating the relationship between EMI integration and sustainable development (SD), we aim to uncover insights that can inform evidence-based policymaking and institutional practice.

Our primary objective is to examine whether a harmonious equilibrium exists between the implementation of EMI policy and its outcomes by analysing the perspectives of both teachers and students. This involves a critical examination of the current landscape of EMI, which emerged from consolidating four main themes: the drivers of EMI implementation, the current status of EMI integration, strategies for managing English language usage and content comprehension within EMI settings, and the challenges hindering the effective implementation of EMI in Algerian HEIs. The analysis, interpretation, and discussion of these main themes and their respective sub-themes in the subsequent subsections enable us to offer a comprehensive portrayal of the contemporary landscape of EMI within Algerian HEIs. Through this exploration, we aim to contribute to the discourse on educational reform and SD, a topic that has not been previously addressed, especially in relation to Algeria.

5.5.1 EMI Implementation Drivers

In this theme, we discuss the nuanced motivations propelling the adoption of EMI in Algerian higher education, as perceived by various participants. Within the contemporary landscape of EMI implementation in Algeria, two drivers emerge, evoking parallels with the pivotal decision in the 1990s to integrate English into primary education. These motivations not only reflect the contemporary needs of globalisation but also resonate with the enduring sociocultural narratives shaped by Algeria's complex colonial history with France. As we navigate through these intertwined threads of past and present, our exploration aims to illuminate the multifaceted dynamics underpinning the trajectory of EMI in Algerian higher education.

Amidst the chorus of condemnation directed at the educational system, calls for the adoption of EMI resonated strongly. For many, the proposal to embrace English represented not only a departure from the legacies of colonialism but also a pragmatic pathway to accessing global opportunities. It was seen as a means to break free from the confines of a linguistic legacy inherited from the colonial era and to position Algeria on a more competitive footing in the international arena. Moreover, proponents of the shift to English in HE highlighted the potential for enhanced international employment prospects. They argued that the previous government had overlooked the importance of English proficiency in securing lucrative employment opportunities on the global stage. By advocating for the incorporation of English into the educational curriculum, they sought to equip Algerian students with the linguistic skills needed to navigate an increasingly interconnected and competitive job market.

5.5.1.1 *Decolonising Education and Scientific Research*

Amidst the longstanding debate over the prominence of Arabic versus French in Algeria, the events of the 2019 protests unveiled a striking readiness among Algerians to embrace the English language. Enduring decades of French hegemony across various sectors, Algerians demonstrated a palpable eagerness to reshape the linguistic landscape, viewing the adoption of English as a means of liberation. The shift in focus towards English in education was seen as a decisive step towards breaking free from the linguistic shackles of the colonial past and asserting a newfound sense of global belonging and engagement. Amidst the calls for reform during the protests, demands for English as a tool for decolonisation in Algeria reverberated prominently (Jacob, 2020). The subsequent formation of a new government cabinet in December 2019 provided an opportune moment for advancing these language demands, signalling a potential shift in official attitudes towards linguistic diversity and inclusion. Notably, Tayeb Bouzid, the former Minister of Higher Education, spearheaded efforts to explore the integration of English into university instruction, exemplified by an online referendum aimed at gauging public sentiment on its inclusion as a medium of instruction (Bouherar & Ghafsi, 2021). This pivotal moment underscored the evolving dynamics within Algerian society and governance, with the recognition of EMI representing a tangible response to the aspirations articulated by protestors. As Algeria navigates this transformative juncture, the incorporation of English into HE emerges not only as a linguistic shift but also as a symbol of broader socio-political change and the quest for a more inclusive and globally connected educational landscape.

The introduction of English into Algerian education brings to the forefront the intricate issue of language prestige, which has long been intertwined with the dominance of the French language in Algeria. The association of prestige with French language usage in Algeria has created a social hierarchy wherein speakers of French are often perceived to possess greater social, cultural, and economic advantages, often at the expense of others. However, the emergence of English as a contender challenges this established hierarchy and disrupts the notion of language prestige associated with French. The discourse surrounding language in Algeria has been instrumental in shaping political ideologies and orientations (Jacob, 2020). Over time, this discourse has perpetuated a dichotomy between Francophones and Arabophones, further reinforcing existing societal divides. With the introduction of English, however, this dichotomy is being challenged, as English represents a departure from the linguistic legacies of French colonialism and introduces new connotations of language that are seen as more aligned with notions of liberation, freedom, and globalisation. While English also carries its own imperial connotations, these have yet to be as damaging as those associated with the French language. The introduction of English thus signals a shift in the linguistic landscape of Algeria and opens up new possibilities for redefining linguistic identities and challenging entrenched power dynamics. As Algeria grapples with these complexities, the role of language in shaping social, cultural, and political relations continues to evolve, with English emerging as a potent force for change and reimagining linguistic ideologies in the country.

The erosion of the French language's dominance in Algeria is indeed taking various forms, particularly within the sphere of higher education. With the introduction of EMI in higher education, concerted efforts are made to encourage teachers to adopt English as the language of scientific research and publishing, effectively relegating French to a secondary or tertiary position in academic production. This shift reveals the growing recognition of English as a neutral language devoid of the social and political ideologies associated with French, while also acknowledging its global standing as the language of research. As such, the adoption of English signifies not only visibility but also enhanced readability of scientific output from Algerian universities on the global stage. However, questions arise about the composition of leadership within HEIs, where Francophone proponents still wield significant influence. One of the participants, Afaf, seems to touch on that clearly in her department claiming that "we still use French, we correspond officially in French, and teachers communicate in French/Arabic, not hint of English whatsoever". Ahmed also shares similar concerns about the influence of Francophones in decision-making roles within HE, claiming that "yet, this might confront the struggles of the Francophone majority in the schooling system as in higher education". Despite the purported shift towards English, the continued presence of Francophone administrators raises doubts about the seriousness of efforts to transition away from French dominance. It seems incongruous to claim the significant role of English while retaining or even promoting individuals who uphold Francophone-centric ideologies within academic institutions. Furthermore, while there is a stated intention of "defrenchification" (in Afaf's words) in higher education, it is notable that more attention is given to training educators than students. This asymmetry in focus adds to shake the government intentions to decolonise HE, suggesting a potential gap in the implementation of language policies. Amel, one of the participants, claims that in order to foster the EMI in higher education, we "require scrutiny and clarity and thoughtful strategies far from superficial, impromptu decisions, with the necessity of seeking the assistance of experts in this field, strengthening trust among stakeholders, and taking into account the issue of resistance to change". While her ideas may suggest a reluctance to embrace change, they also emphasise the significance of English in mitigating language conflicts in Algeria and opening up new global employment prospects.

5.5.1.2 Unlocking International Opportunities and Job Market Prospects

The adoption of EMI in Algerian HE is perceived by participants as one of the drives towards addressing longstanding disparities in the job market, both locally and globally. Algeria's reliance on French as the dominant language in professional settings has marginalised ambitious job seekers who are not fluent in French (Alkhateeb & Bouherar, 2022). The job market has been heavily biased towards hiring French speakers, leaving graduates proficient in Arabic or

English with limited opportunities within the country. Yacine, one of the participants, articulates this sentiment, suggesting that "there could be some opportunities that the French language did not provide. … It could be that the university is looking to promote competition in its local job market". He emphasises the potential for English to introduce competition into the local job market, thereby expanding opportunities for all graduates. Amina echoes these sentiments, highlighting how EMI facilitates internationalisation and equips Algerian students with the skills needed to compete on a global scale, claiming that "EMI facilitates internationalisation and enables Algerian students to compete in the global job market". The consensus among participants is that proficiency in English can significantly enhance graduates' prospects in the global job market, where English is widely used. This aligns with the broader trend of globalisation, where proficiency in English has become increasingly valuable for accessing diverse employment opportunities. Moreover, the implementation of English in primary schools has already begun to yield tangible benefits in terms of job creation. The hiring of 5000 English teachers to support the introduction of English as a new subject in primary education (The New Arab, 2022) exemplifies the potential for EMI to stimulate job growth at the national level. This initiative not only addresses the immediate need for English language instruction but also lays the foundation for future employment opportunities in various sectors. Sectors like economy can also benefit from EMI with the example of multinational companies operating in Algeria that are increasingly prioritising employees with strong English proficiency and employment skills. By recruiting locally trained professionals with English language competence, these companies can reduce costs associated with hiring expatriates while simultaneously benefiting from a highly skilled workforce.

While participants' arguments regarding the importance of English in Algerian HE are legitimate, it is important to recognise that their openness to English also reflects a longstanding resistance to the dominance of French in Algeria. The emergence of the Al Hirak protest movement has played a pivotal role in elevating the status of English in education, signalling its potential to serve as a tidal wave in Algeria (Maraf & Vanci Osam, 2022). However, despite the growing acknowledgement of English's importance, the notion of it becoming a tidal wave in Algeria, as suggested by Maraf and Vanci Osam, may be subject to scrutiny, particularly in light of the resistance that EMI continues to face in higher education.

Indeed, while English may offer a promising alternative to French, its adoption also raises concerns about the potential marginalisation of national languages such as Arabic and Tamazight. As Bentahar (2020) points out, there are cultural and linguistic complexities surrounding the promotion of Derja (vernacular Algerian Arabic) as a means of resistance against the spread of both Arabic and French, since Al Hirak. Thus, while English may present opportunities for linguistic diversity, it also poses challenges to the preservation of Algeria's linguistic heritage. Amidst this linguistic complexity, English has the potential to promote employment equality and opportunity in Algeria. However, its broader aim may also include replacing French dominance with English. This is evident in the push to compel educators in social sciences, where Arabic serves as the medium of instruction, to transition to English for scientific production and future teaching instruction. While this shift may align with efforts to internationalise Algerian HE and increase graduates' competitiveness in the global job market, it also reveals the tension between linguistic diversity and the hegemony of English language. In navigating these complexities, it is essential for Algerian policymakers and educators to approach the promotion of English with sensitivity and caution. Efforts to integrate English into HE should be balanced with measures to preserve and promote linguistic diversity, ensuring that national languages are not marginalised in the process, as Amel claims, "the English language must be a complement to the Arabic language and not a substitute for it, because this will further exacerbate our linguistic problems and fuel linguistic pollution even more". In fact, the task of language policymakers in Algeria is a delicate balancing act. On one hand, there is a pressing need to strengthen the job market by promoting English proficiency and enhancing graduates' competitiveness on a global scale. On the other hand, any language policies implemented must be mindful of the complex linguistic landscape of Algeria, with its predefined priorities and aims. Language policymakers must navigate a myriad of considerations, including the preservation of linguistic diversity, the promotion of national languages, and the integration of global languages such as English. Any policies set forth must seek to fortify the job market without exacerbating existing tensions or marginalising certain linguistic communities.

5.5.2 *EMI Integration Status*

The reactions of participants to the introduction of EMI in HE reflect the early stages of its implementation and the complexities involved. The acknowledgement of decolonising aims behind EMI shows the entrenched dominance of French in the educational sphere for a considerable period. This recognition implies that the integration of EMI is likely to encounter obstacles and uncertainties. Critics may argue that while EMI may face initial challenges, its efficacy can improve over time as institutions gain unprecedented experience with its implementation. However, decision-makers must carefully consider the nuances surrounding this decision. This includes addressing the concerns of teachers who have extensive experience with French as a medium of instruction, students who are accustomed to learning in French, and curricula that are predominantly French based. Replacing these factors in a short time frame without proper planning is likely to be challenging and may require years of careful consideration and adaptation. It is essential for decision-makers to develop comprehensive strategies that take into account the diverse needs and perspectives of stakeholders and ensure a smooth transition to EMI. By adopting a thoughtful and inclusive approach, institutions can navigate the complexities of integrating EMI into HE and ultimately enhance the quality and accessibility of education in Algeria.

5.5.2.1 EMI in Its Nascent Stage

One early attempt to integrate English into Algerian education dates back to the 1990s when it was proposed as an additional subject alongside French in primary schools. However, this initiative was ultimately abandoned, resulting in the resignation of the Minister of Education at the time (Bouherar & Ghafsi, 2021). Despite its failure, this endeavour raised significant questions about the future role of English in Algerian education. A second attempt was made following the inauguration of a new government, succeeding the administration of Bouteflika. The momentum for integrating English gained traction, culminating in the first practical sign of EMI in HE during the 2023/2024 academic year. This milestone was preceded by a series of deliberations, decrees, and meetings aimed at laying the groundwork for the implementation of EMI. Key decrees issued on 12 February 2023 (No. 0005 and No. 0006) directed university chancellors to establish pedagogical committees to monitor the progress of English implementation within their institutions. Additionally, a national committee comprising 18 members was established to oversee the introduction of English across public universities in Algeria and supervise the activities of local committees. Furthermore, decree No. 107, dated 22 March 2023, outlined the appointment of teachers responsible for curriculum design and IT engineers to provide support to educators throughout the implementation process.

The decrees issued by the Ministry reveal a clear emphasis on training courses for teachers, indicating a recognition of the pivotal role educators play in the successful implementation of EMI in Algerian higher education. However, while the focus on teacher training is crucial, it also highlights potential challenges and areas for further consideration. According to Yacine:

> I can see that the Ministry started to fail toward promoting English in the university without proper training programmes, especially for students. There is a typical picture that students that are originally from big cities would not face difficulty in English and therefore assuming that perhaps, students do not need training, which is wrong.

The stereotype of metropolitan students as proficient foreign language speakers reflects the mentality of decision-makers, as highlighted by Bouherar and Ghafsi (2021), where approximately 80% of first-year students fail exams due to their lack of proficiency in French, particularly among non-Algiers residents. With some teachers utilising EMI in their classrooms, students who lack sufficient English proficiency are left struggling to comprehend complex concepts. Ahlem claims: "[S]tudents struggle to understand different concepts, especially the complicated ones because of the language".

Indeed, participants seek multiple justifications for the challenges EMI faces in Algerian higher education, recognising the nascent nature of its implementation. Nessma's words offer a cogent rationale for these challenges as she claims:

> The current status of English as a medium of instruction in Algeria appears to be in the early stages of implementation, with efforts underway to integrate it into higher education institutions. While progress has been made, challenges such as infrastructure development, faculty training, curriculum design, and language proficiency among students and instructors may still exist. Ongoing evaluation and adaptation are likely needed to ensure the effective implementation and long-term success of EMI in Algerian higher education.

Therefore, is it logical to encounter such challenges when implementing EMI for the first time? Well, any individual using logical reasoning would argue that this is the natural progression for EMI to develop and generate improved results in HE by addressing existing shortcomings. However, it is a concern for many others who perceive these challenges as not reflecting a strong political will, such as Nour who contends: "[T]he Algerian educational system is already suffering from endless sudden decisions to incorporate new policies that are being put forward without adequate need and situational analysis". Similarly, Kenza voices a similar concern to Nour, suggesting: "[I]nconsistent policies, regulations, and governance structures may hinder the effective implementation of EMI across institutions". Nevertheless, as previously suggested in this book, while the government appears to approach its policies in a tokenistic manner, it is unreasonable yet to determine whether the unprecedented experience of implementing EMI in HE required careful planning that was not promptly executed, or whether the government's tokenistic policies are too premature to assert such a statement. What is evident is that one year has passed since the official practical implementation of EMI, yet the HE sector still predominantly uses French as the medium of instruction.

5.5.2.2 Difficulties in Transitioning Between French and Arabic to English

Amidst these intricacies, students find themselves grappling with numerous linguistic challenges stemming from the integration of English and its attendant learning implications. The nascent phase of English implementation has introduced certain instabilities regarding the utilisation of other languages, namely French, Arabic, and occasionally English. Consequently, three distinct groups have emerged: Arabic-speaking students, who are inclined to utilise Arabic as their primary mode of communication; French-speaking students, who predominantly engage with French as the language of instruction; and English-speaking students, who favour the use of English in classroom settings. The potential dilemma lies in the presence of a mixed-class environment comprising all three groups. In such scenarios, irrespective of the language of instruction employed, at least two groups may feel entirely marginalised from the learning process.

Kormos (2017) outlines how cognitive and psychological differences of students may hinder the learning abilities of the same second language. In the Algerian case, we are talking about how students of different linguistic and maybe cultural backgrounds come to struggle to understand a language they have less competence in. Therefore, two tasks await students: one is to understand the language being used in class and second is to grasp the lessons delivered through English. This dual challenge faced by students shows the multifaceted nature of language learning and its intersection with cognitive and psychological factors, let alone linguistic and cultural ones. Not only must students navigate the complexities of learning a second/third language, but they must also contend with the added layer of comprehending subject matter presented in that language. This becomes particularly daunting in the Algerian context, where students hail from diverse linguistic and cultural backgrounds, each bringing their own set of experiences and competencies to the learning environment. Moreover, the issue is compounded by the dearth of adequate training and support mechanisms available to students. Without sufficient preparation and guidance, students may find themselves ill-equipped to navigate the linguistic and cognitive demands of learning through the English language. Ahlem asserts:

> Students, on the other hand, face many difficulties in understanding the subject because of their limited knowledge of the language. They also struggle to explain well using English even though they have the information, resulting in bad grades and low performance.

Resorting to Arabic or French as a means of decoding the language used in class is a common strategy employed by students in the hopes of better understanding the lesson being delivered. In line with that, the study conducted by Bälter et al. (2023) indicates that the introduction of EMI in Swedish HE has adverse effects on students' academic performance, with students appearing to learn less effectively when EMI is utilised. Despite the limited research on the negative outcomes of EMI on academic performance (Macaro, 2018), the push to implement English in Algeria is taking various forms and gaining momentum.

Students are not the only affected population in HE from the negative outcomes of EMI in academia, but teachers are also concerned. Teachers also seem to struggle using EMI in their classes due to the insufficient linguistic competencies and it would as a consequence influence their students' academic performance. Farah claims that "teachers who did not study English as a field of speciality will find it so difficult to teach in English". The adverse effects of EMI in academia extend beyond students to also impact teachers in higher education. Teachers also often grapple with the challenges of implementing EMI in their classes, primarily due to insufficient linguistic competencies. This struggle not only affects their ability to effectively communicate and deliver lessons in English but also has repercussions on their students' academic performance. Indeed, the linguistic demands of teaching in English can present significant hurdles for educators who may lack formal training or proficiency in the language. Without adequate preparation and support, teachers may struggle to convey complex concepts and facilitate meaningful interactions in the classroom.

This raises important questions about the potential implications of EMI implementation in Algerian HE. While the adoption of English may be driven by aspirations for internationalisation and enhanced global competitiveness, it is essential to carefully consider its impact on students' learning outcomes. Moreover, the findings from Bälter et al.'s study highlight the need for further research to explore the nuanced dynamics at play and to inform evidence-based decision-making in the realm of language education policy. As Algeria navigates this transition, it must prioritise a balanced approach that addresses both the opportunities and the challenges associated with the integration of EMI.

5.5.3 *Managing English Utilisation and Content Comprehension in EMI Settings*

English has gained global prominence partly due to globalisation and internationalisation, leading to the widespread adoption of EMI in HEIs worldwide. Recently, Algeria has introduced EMI, recognising English as a key language for development. However, EMI is most effective in contexts where both teachers and students are proficient in English, and where teachers are trained to deliver high-quality instruction in this language, as EMI focuses on content instruction without explicit language learning objectives (Dafouz, 2018). To support this transition, training programmes were organised to enhance teachers' language skills so that they could teach their subjects in English. Nevertheless, these programmes primarily improved lecturers' language proficiency without addressing their pedagogical skills, which are crucial for maintaining educational quality. As a result, these lecturers may not be fully prepared to transition to EMI in their classes. Furthermore, many students, especially in scientific disciplines, do not possess adequate levels of English proficiency. In Algeria, English is not a second language and is scarcely used outside the classroom. EMI classes focus on teaching content through English rather than teaching English itself. Assuming that students have high levels of English proficiency risks undermining their mastery of their respective disciplines.

CLIL, which addresses both content and language learning, seems more suitable for the Algerian context than EMI, which primarily focuses on content learning. However, Algeria has adopted EMI with the expectation that teachers will perform dual roles: teaching both content and English, so that students graduate with proficient knowledge of their disciplines and English. While students can acquire English through its application to their specialities, as language is best learnt when applied to relevant subjects (Brinton et al., 2003), EMI classes primarily aim to teach subjects without explicit language instruction. Therefore, students lacking good English proficiency will struggle with both language acquisition and content comprehension. The interviewees of this study expressed concerns about how lecturers could use English without compromising content comprehension in EMI classes. Students are supposed to benefit from EMI classes when they are proficient in English. Those without sufficient English proficiency face two problems: language proficiency and content comprehension. Students who are not proficient in English may perceive themselves as lacking proficiency in their disciplines, leading to demotivation and discouragement from acquiring both English and content knowledge (Carrió-Pastor, 2021). Therefore, the inclusion of EMI in Algerian HEIs might not foster equitable and quality education, which is essential for providing lifelong learning opportunities for all. This contradicts SDG4, suggesting that the recent reform of introducing EMI in Algeria cannot be sustained without effectively managing the integration of English and content learning.

The theme of EMI and the comprehension of content in HEIs, as emerged from the interviews, have been under-researched in previous studies within the field. Lin and Lei (2021) examined the effects of EMI on students' academic outcomes. Their findings revealed that students' proficiency in English significantly influenced their academic performance, suggesting that the language barrier to academic success can be mitigated through effective language support and strong academic capabilities. This indicates that students with high English proficiency achieve better academic outcomes compared to those with lower proficiency. It is, therefore, crucial for educators to consider these differences in their teaching practices to attain desirable outcomes. This conclusion was also reached by Phuong and Nguyen (2019) in their exploration of university students' perceptions in Vietnam, where English is neither a second language nor widely used outside the classroom, regarding the advantages and disadvantages of EMI classes. Their data analysis indicated that students viewed EMI lessons positively in terms of resource availability, enhanced English skills, and future job prospects. However, they noted that low English proficiency and insufficient vocabulary knowledge negatively impacted students' achievements in EMI classrooms. This suggests that while EMI in specialised subjects can enhance students' language and content learning, adequate vocabulary size and proficiency are necessary to maximise the benefits from EMI lessons. Both teachers and students in EMI classes possess varying levels of English proficiency, which can lead to differing levels of content comprehension. Lecturers should be cognisant of this factor and incorporate it into their content delivery.

Siegel (2020) conducted a study on comprehension in Swedish EMI lectures and the impact of lecturers' use of English. The findings showed that lecturers and students do not have equal proficiency, resulting in a wide range of success and learning outcomes in EMI settings. Factors such as English listening ability, background knowledge of the subject, and motivation must be managed for successful EMI lectures. Tran et al. (2020) similarly found that instructors' teaching methods, along with both students' and instructors' language competence, hinder students' understanding of lessons at a university in Vietnam. In line with these studies, some interviewees in this study also noted the difficulty of using EMI while ensuring the understanding of content because of lecturers' and students' poor English and vocabulary knowledge. For example, Ryad said, "[I]t is hard for teachers who do not have the required English language skills". Fairouz also mentioned that implementing EMI would increase the difficulty for students who would struggle to switch from a second language, French, to a foreign language, English, which would lead them to seek other ways to understand the content, like translation. She added:

> It is a little bit difficult to balance between the use of English with ensuring that the students are getting the information properly, especially if the level of the students who used to focus on French as a second language their whole learning journey, I think this is the most challenging obstacle, students will struggle twice, instead of focusing their efforts on getting the knowledge, they will find themselves translating also

According to the perspectives of the interviewees, the integration of EMI in Algerian HEIs complicates the tasks of both teachers and students primarily due to the language barrier, which disrupts the balance between English utilisation and content mastery. The majority of respondents proposed strategies such as translanguaging, implementing various materials, and integrating ESP and CLIL to manage the use of English and enhance content comprehension in EMI classes. Each of these sub-themes is discussed in detail below.

5.5.3.1 Leveraging Translanguaging Practices Within EMI Classes

The integration of EMI into HEIs can be pursued through either an English-only policy, aimed at enhancing students' English proficiency, or a flexible bilingual or multilingual policy, designed to improve both learning and overall well-being. In Algeria, the current EMI approach appears to favour an English-only policy, with lecturers encouraged to teach exclusively in English due to its global importance. However, this strategy seems to overlook the significant role of Algeria's linguistic diversity, potentially neglecting the benefits of incorporating multilingualism into educational practices. Although Algerian stakeholders typically do not advocate for a multilingual approach when discussing EMI, lecturers may employ such strategies to facilitate content comprehension and improve students' English proficiency. Given their limited proficiency in English and the longstanding use of French as a medium of instruction, both lecturers and students are likely to draw on their linguistic repertoire, including French, Arabic, or even Tamazight, to navigate learning. This dynamic process, in which multilingual learners utilise multiple languages as an integrated system for effective communication and understanding, is known as "translanguaging" (Canagarajah, 2011). While participants did not explicitly mention this concept, likely due to their lack of awareness, their responses reflect its use as a strategy to balance the use of English and content learning.

Most interviewees indicated that incorporating translanguaging in EMI classes would enhance content comprehension and English language acquisition. For instance, Ahlem noted, "Educators need to use different languages to explain different concepts. This would help learners develop their vocabulary and general understanding of the subject". Students shared similar perspectives; Taki mentioned that incorporating Arabic alongside English provides linguistic support and adapts the curriculum to accommodate students' proficiency levels. As in previous studies (Branka & Molino, 2022; Smit, 2019; Van der Walt, 2013), the interviewees suggested that translanguaging serves various pedagogical purposes, including explanation, interaction, clarification, understanding, and accommodation. Hanane stated, "Using all languages as resources in the classroom", while Amel noted that translanguaging can facilitate teachers' delivery of their lectures and enhance their English language proficiency. Zohra added that translanguaging helps students and lecturers feel more at ease with EMI, gradually increasing the use of English as comfort levels rise. These findings align with those of Yilmaz et al. (2024), who found that using multilingualism as a resource in EMI classes at a Korean university is productive, aiding students in constructing, promoting, and demonstrating learning, fostering compassion, and enhancing understanding of both their home and new languages.

One lecturer noted that rigid teachers who exclusively use EMI in their sessions hinder students' understanding and success in their modules. Yacine observed that some students have complained about the inflexible attitudes of certain educators, stating, "[R]igid mentalities of some educators saying that they use EMI in class and therefore this affected our comprehension of what is going on in the class and marks as a result". Withholding students from the opportunity to use their linguistic repertoire to understand content and learn English violates principles of social justice and access to impactful learning (Van der Walt, 2013). This approach contradicts the SDG4, particularly target 4.3, which emphasises equal access to education, and target 4.5, which seeks to eliminate all forms of discrimination in education. Embracing a multilingual instructional approach not only addresses diverse linguistic backgrounds but also fosters inclusivity within the academic setting. This inclusive pedagogical model recognises linguistic diversity as a resource that enriches learning experiences. Implementing a multilingual policy can enhance students' language skills beyond English, promoting broader communication competencies essential for success in a globalised world.

This study has implications for policymakers, highlighting the need to reconsider top-down policies by incorporating bottom-up perspectives and recognising multilingual approaches as valuable resources in EMI classes. It also provides recommendations for EMI practitioners to allow the use of multiple languages and adopt translanguaging as an efficient strategy for managing English utilisation and content comprehension, ultimately leading to better learning outcomes and increased awareness of the multilingual world. This would contribute to SDG4, particularly target 4.7, which promotes SD and global citizenship (GC) through inclusion, a culture of peace, and an appreciation of diversity.

5.5.3.2 Incorporating Diverse Materials in EMI Settings

In addition to translanguaging, the interviewees emphasised the importance of integrating diverse materials to enhance comprehension in EMI classes. While the impact of translanguaging on content and language learning in EMI settings has garnered some attention in academic literature, a cursory Google search suggests limited exploration of integrating various materials to effectively manage content and support EMI usage. This research addresses this gap by examining the perspectives of Algerian university educators and students, highlighting the potential of employing a variety of instructional materials to improve comprehension of both content and language in EMI classes. The findings of this study provide valuable insights for enhancing educational outcomes in multilingual environments.

Some interviewees believed that using different aids would make the content of the EMI class accessible to all students regardless of their levels of proficiency. They highlighted the importance of using diverse teaching materials. For instance, Amina mentioned, "using visual aids and other clues to make the content more accessible for students". They emphasised the need for teachers to prepare comprehensive materials, with Wafa stating that "teachers should prepare holistic materials". Nessma provided examples of visual aids that could be incorporated in EMI classes to improve comprehension, such as diagrams, charts, vocabulary lists, and multimedia. She believed that such materials are flexible and can be adapted to students' language proficiency levels and learning needs. Similarly, Ahlem suggested that enhancing classroom practice with various materials would improve outcomes, particularly in students' English and content learning. She noted, "Their teaching strategies can be supported by videos, diagrams, charts, figures, etc. This would help the learners develop their vocabulary and general understanding of the subject". Kenza added that a variety of materials would reinforce understanding, assist in assessment, and provide constructive feedback to guide learning. Zahra emphasised the importance of using clear and comprehensible language and avoiding complex vocabulary and structures that could confuse learners. This perspective was endorsed by a student, Yasmine, who stressed the importance of using materials with simple language to support learning. She remarked, "[S]implifying language in explanations and materials can make complex concepts more accessible to all students, even those with low proficiency in English".

These findings suggest that employing diverse and adaptable teaching materials can make EMI classes more accessible to students with varying levels of English proficiency. Visual aids and multimedia are particularly effective in enhancing comprehension. The interviewees emphasised the importance of preparing comprehensive and flexible materials, alongside using clear and simple language to avoid confusion and make complex concepts more understandable. Incorporating these elements into EMI classes is believed to improve both language and content learning, provide constructive feedback, and ultimately enhance educational outcomes. Im and Kim (2015) also observed that blended learning enhances the learning process and outcomes in EMI classes, rendering it an effective strategy for implementing EMI in non-English-speaking HEIs. Banegas and Busleimán (2021) recommended that EMI teachers develop context-sensitive materials incorporating glocalisation, include activities designed to enhance learners' language awareness with specific vocabulary and textual grammar, ensure that the materials used align with the affordances, possibilities, and limitations of the learning environment, and promote autonomous learning. The analysis of data suggests that incorporating a variety of materials in EMI classes enhances the educational experience, benefiting both Algerian university teachers and students. This strategy not only enhances the quality of education but also fosters SD and promotes a sense of GC.

5.5.3.3 ESP and CLIL in EMI Classes

As elaborated in Sect. 5.3, ESP, CLIL, and EMI are distinct approaches, each with a different emphasis on language and content, but all are rooted in CBI. ESP is primarily language-focused, catering to the specific linguistic needs of learners in particular professional or academic contexts. CLIL, on the other hand, places equal importance on content and language, aiming to teach subjects through a foreign language, thereby promoting both subject matter understanding and language proficiency. EMI, being more content-focused, involves teaching academic subjects exclusively in English without explicit language instruction, assuming that language acquisition will occur incidentally through content engagement (Airey, 2016). Several interviewees suggested that both ESP and CLIL could serve as useful strategies to support content delivery and facilitate English learning in EMI classes.

They indicated that the specialised language support provided by ESP can render complex subject matter more accessible to learners who may struggle with the linguistic demands of EMI. Fatima asserted that ESP would enhance students' speaking skills, thus necessitating its incorporation into the learning materials in EMI classes: "[E]nsuring students' complete ability and control of speaking English by providing them with ESP (English for Specific Purposes) along with their usual learning material". Abderrezzaq shared a similar perspective on integrating ESP, adding that students should take the initiative to improve their language skills by attending language centres and working on their comprehension skills: "I suggest introducing and focusing on ESP. This is how teachers can ensure understanding of content and use of English. Students should also be encouraged to join intensive language centres and work on their comprehension skills". Yasmine, a student, also advised her teachers to provide ESP, which would enhance their field-specific vocabulary and grammar: "Offering English for Specific Purposes (ESP). These courses can equip students with the specialised vocabulary and grammatical structures relevant to their field of study". These findings indicate that ESP can make EMI sustainable by promoting inclusive and equitable quality education, which is the basis of SDG4.

A few interviewees highlighted the advantages of CLIL in fostering a more integrated comprehension of both content and language. They observed that

CLIL's dual focus enables students to concurrently enhance their language skills while deeply engaging with the subject matter, resulting in more comprehensive learning outcomes. Nezha, for instance, mentioned that ESP and CLIL are useful for managing content and language understanding: "[E]ducators in Algeria (or anywhere with EMI) can balance English language development with ensuring students grasp content effectively through teaching specific linguistic repertoire, content and language integrated learning". Nihed also stated that "[c]ontent-language integration" is an effective strategy in EMI sessions. Although EMI and CLIL are distinct approaches (see Sect. 5.3), the interviewees pointed out that CLIL is a versatile strategy that places equal emphasis on English acquisition and content comprehension in EMI classes.

The findings indicate that the specialised linguistic support provided by ESP can effectively address the specific language challenges encountered by students. Simultaneously, the CLIL approach can enhance overall comprehension and retention of subject content. Consequently, the integration of both ESP and CLIL into EMI classes could bridge the gap between language proficiency and academic performance, thereby offering a more holistic educational experience. Costa and Mastellotto (2022) contend that ESP within EMI classes equips students for communication, develops their academic disciplinary and transversal skills, and fosters intercultural communication. Therefore, the integration of CLIL and ESP within EMI establishes a foundation for effective learning and the acquisition of pertinent vocational and technical skills, thereby promoting decent work and GC in a reconnected world, which are targets encapsulated in SDG4.

5.5.4 *Obstacles to Adopting EMI in Algerian Higher Education*

Among the drivers for establishing an EMI policy within Algerian higher education, as outlined in this study (see Sect. 5.4.1), is fostering development by enhancing research visibility and the internationalisation of higher education. Nevertheless, participants have voiced that the execution of the EMI policy to attain these objectives is impeded by various challenges, encompassing inadequate resources, linguistic marginalisation, and insufficient professional development for educators.

Previous scholarship in this domain has delineated the hurdles to implementing EMI in HEIs, yet this issue remains unaddressed in Algeria. For instance, Vu and Burns (2014) observed that Vietnamese lecturers encountered obstacles concerning linguistic proficiency, pedagogical issues, and the availability of resources. This indicates a broader, systemic issue transcending individual capabilities to encompass institutional and infrastructural support systems. Similarly, Kaur (2020) noted that educators and students at a Malaysian public university grappled with difficulties in EMI classes, such as student passivity, limited comprehension of peer contributions during interactions, and the practice of code-switching to the majority's primary language, thereby alienating international students. These findings underscore the intricate interplay between language policy and classroom dynamics, suggesting that merely altering the medium of instruction does not automatically translate into improved educational outcomes without addressing underlying pedagogical and linguistic concerns. Alhassan (2021) discovered that EMI lecturers in Omani HE confronted linguistic and pedagogical challenges, underscoring the necessity for teachers' training and professional development. This accentuates the importance of comprehensive support systems encompassing continuous professional development opportunities for educators, ensuring they possess the requisite skills and knowledge to effectively disseminate content in a foreign language.

Drawing upon the findings of this study, several recommendations are offered to inform policymaking and enhance the efficacy of EMI implementation and its outcomes in Algeria. Firstly, it is imperative to furnish the requisite resources, including teaching materials, technological tools, and institutional support, to cultivate an enabling environment for EMI. Secondly, emphasis should be placed on providing linguistic support to mitigate marginalisation in the classroom. This may entail language enhancement programmes for both students and educators to enhance their proficiency and confidence in utilising EMI. It could also involve raising awareness about the importance of employing translanguaging practices to foster comprehension. Lastly, it is crucial to offer targeted training for educators, focusing on efficacious pedagogical strategies for integrating EMI in content classes. Such training should encompass areas such as content delivery, student engagement, and assessment techniques tailored to EMI contexts. The implementation of these recommendations necessitates a collaborative approach involving policymakers, educational institutions, teachers, and students. By addressing these challenges holistically, it is conceivable to realise the dual objectives of increasing research visibility and internationalisation of HE while fostering an inclusive and supportive learning environment.

5.5.4.1 *Inadequate Resources*

The interviewees highlighted that the incorporation of diverse, comprehensible materials in EMI classes significantly enhances both language acquisition and understanding of the subject matter (see Sect. 5.4.3.2). This result aligns with the findings of prior researchers who have contended that the utilisation of various resources in EMI contexts facilitates the overall teaching and learning process, as well as the outcomes achieved (Banegas & Busleimán, 2021; Im & Kim, 2015). However, a recurrent theme among some participants was the issue of "limited resources", which they identified as a major obstacle to the effective implementation of EMI in Algerian higher education.

The term "limited resources" frequently surfaced in interviews, indicating a consensus on the adverse impact of inadequate infrastructure on educational quality. High-quality infrastructure is essential as it creates an environment conducive to learning and engagement, providing the necessary comfort and support for effective instruction. Unfortunately, the HE sector in Algeria is plagued by insufficient resources, rendering the integration of EMI into content classes a formidable challenge for both educators and students. For instance, Ahmed remarked, "Provisions in terms of instructional resources in English might remain the most hindrance", underscoring the critical need for appropriate instructional materials. Participants advocated for better resourcing of universities to enable the effective implementation of EMI. Kenza emphasised, "Limited resources, including instructional materials, technology infrastructure, and funding, may hinder the implementation of EMI programs. This will require increased investment". Similarly, Nessma highlighted the necessity of investing in resources such as textbooks, technological tools, and language support services to facilitate EMI implementation, noting, "Resources and infrastructure: Investing in resources such as textbooks, technology, and language support services to facilitate EMI implementation". These views underscore a shared belief among participants that the policy of integrating EMI into Algerian HE must be accompanied by increased financial support to provide adequate teaching and learning materials, technological aids, and opportunities for language enhancement.

Hanane pointed out the deficiencies at the intensive language training centre, CEIL, which offers language courses for teachers and students, lamenting that the centre is poorly equipped and that its staff are underpaid: "The CEIL centres are not equipped with adequate materials for teaching (sometimes, there are no rooms)". Amel differentiated between two types of challenges: "material and human". Material challenges involve the lack of sufficient pedagogical and digital resources, while human challenges pertain to the limited number of professors and supervisors proficient in English. To address this gap, the Algerian Ministry of Higher Education and Scientific Research is currently expanding the number of doctoral programmes in English for the 2024–2025 academic year. However, these students will require at least four years to complete their PhDs and begin teaching at universities after passing their qualifying exams. Amel highlighted "the challenge of time", indicating that the EMI policy was implemented hastily without thorough preparation.

Traditionally, content classes in social and human sciences were taught in Arabic, and those in scientific fields were conducted in French. Consequently, most of the available resources in university libraries are in French. Souheyla noted that "another challenge is finding sources in English", a sentiment echoed by student Yasmine, who stressed the importance of "providing more resources in English". It is crucial to enhance the libraries' collections with English-language books to support the successful application of EMI in Algerian HEIs. The scarcity of resources significantly hampers the development of high-quality education, which is central to achieving SDG4 within EMI classes. The findings suggest that the implementation of EMI in Algerian HEIs may not be sustainable in the long term due to the insufficient availability of both human and material resources, including essential pedagogical and technological aids.

5.5.4.2 Linguistic Marginalisation

The implementation of EMI in Algerian HEIs has elicited significant concern regarding the potential for linguistic marginalisation. The participants have expressed apprehension that relying solely on English for content delivery risks undermining Algeria's linguistic diversity. This concern extends beyond the official languages—Arabic and Tamazight—to include French, as well as students and educators with limited English proficiency.

For example, Tarek highlighted the necessity for strategic preparation: "setting the ground for such a process with specific kinds of teachers". This quote underscores the importance of adequately training and equipping teachers to handle the transition to EMI. Similarly, Farah suggested a cautious, phased implementation of EMI: "[I]t can be tested for some months or even years, then when it proves its success, it might be generalised to all universities". Her recommendation implies that a gradual, experimental approach could mitigate potential disruptions and validate the efficacy of EMI before its broader adoption. Some interviewees criticised the rapid introduction of educational policies without sufficient analysis of the existing linguistic environment in Algerian universities. Nour remarked: "[T]he Algerian educational system is already suffering from endless sudden decisions to incorporate new policies that are being put forward without adequate need and situational analysis". This statement reflects a broader sentiment that rushed policy changes can destabilise the educational system, particularly when they are not grounded in thorough situational assessments. Abderrezzaq's comment, "[W]e cannot ensure the integration of English overnight", echoes this need for a more measured approach to policy implementation. In terms of practical support, Kenza suggested: "Streamlining policies, providing clear guidelines, and establishing quality assurance mechanisms can help create a supportive policy environment for EMI implementation and monitoring". This insight highlights the necessity for coherent policy frameworks and infrastructural support to facilitate the successful adoption of EMI.

A recurrent issue is the risk of linguistic marginalisation. Teachers and students alike have voiced concerns that prioritising English could diminish the use of Arabic, Tamazight, and French in educational settings. Amina, a student, expressed this concern: "[P]rioritising English may inadvertently lead to marginalisation of other languages". This quote emphasises the inadvertent consequences of an English-centric approach on Algeria's multilingual landscape. Similarly, Yasmine noted, "[T]here is a risk of marginalisation as prioritising English could lead to a decline in the use of Standard Arabic, Tamazight and French". She proposed a multilingual approach to prevent such marginalisation: "To mitigate this, I think it is important to allow students and academics to use other languages alongside English and promote a multilingual inclusive environment in higher education". This approach would preserve linguistic diversity while facilitating the learning of English. Amina reinforced the idea of maintaining native language proficiency alongside English: "ensuring students maintain proficiency in their native language while becoming fluent in English". This perspective supports the view that linguistic diversity should be preserved as a means of enhancing, rather than impeding, English learning.

Kenza highlighted the adverse effects of a monolingual approach, particularly for speakers of Arabic, Tamazight, and French, and those with limited English proficiency: "EMI may inadvertently disadvantage students who are not proficient in English or who come from linguistic backgrounds other than English (Berberophone, Arabophone and Francophone)". This quote points to the inequities that an English-only policy could perpetuate, potentially disadvantaging non-English speakers. Van der Walt (2013) supports this view, arguing that an English-only policy in a multilingual setting violates principles of social justice, access to knowledge, and effective learning. Kenza further suggested: creating an inclusive language policy that recognises the linguistic diversity of the educational setting, offering language support programmes, implementing multilingual education models and providing professional development opportunities for enabling educators to teach multilingual classrooms. These recommendations call for a comprehensive policy framework that supports multilingual education and teacher development.

The concept of translanguaging, as discussed by Branka and Molino (2022), Le Hoang Ngo (2021), Smit (2019), and Van der Walt (2013), offers a valuable model for multilingual learners, facilitating both language acquisition and content understanding. Participants called for a bottom-up approach to policy development, one that reflects the linguistic realities of classrooms rather than imposing top-down directives. Abderrezzaq disapproved of language policies driven by political motives: "I am against turning languages into political tools". He advocated for allowing language choices to be determined by the needs of teachers and students: "[N]o language should be prioritised. It is up to teachers and students". Nezha echoed this sentiment, emphasising the importance of embracing linguistic diversity: "[W]e accept linguistic diversity rather than resistance". The anticipated resistance from older staff to changing the medium of instruction was a particular concern. Souheyla noted: "[I]t would be perplexing to expect a teacher in his 50's, who has dedicated 30 years of his life to teaching content in French, to suddenly transition to English as he approaches retirement". This observation underscores the challenges of expecting seasoned educators to swiftly adapt to a new language of instruction. A multilingual approach could be especially beneficial for such educators, enabling them to use multiple languages to facilitate both English and content learning. This approach also helps those with limited English proficiency to avoid marginalisation. Yacine highlighted this risk: "[T]here is a great risk of marginalisation. I just mentioned the cases of teachers and students who still do not have a profound linguistic background that permits them to use EMI". Similarly, Nessma pointed out: "Prioritising English as the medium of instruction in Algerian higher education carries the risk of marginalisation for students and faculty not proficient in English".

Many participants believed that a multilingual approach in EMI settings helps preserve individual identities and fosters a culture of peace. Wafa articulated this by saying: “This language will widen up the marginalisation of Arabic not only French, and the mother language is one of the essential cores in building someone’s intelligence, identity and maintenance in his land”. Hanane added: “Arabic, English, Tamazight and French must be preserved and developed inside the classrooms equally. This contributes to the preservation of identity and social peace”. Taki, a student, supported fostering multilingualism: “fostering the use of multilingualism”. Adopting a multilingual approach in the implementation of EMI in Algerian HEIs can significantly reduce linguistic marginalisation and create an inclusive educational environment that values both linguistic and cultural diversity. This approach aligns with SDG4, particularly target 4.7, which emphasises inclusive and equitable quality education.

5.5.4.3 Insufficient Professional Development for Teachers

The interviewees highlighted the inadequacy of professional development opportunities for teachers as a significant barrier to the successful adoption of EMI within Algerian HEIs. They observed that existing training programmes primarily concentrate on enhancing the language proficiency of the staff, often overlooking crucial pedagogical aspects such as content delivery, student engagement, and assessment strategies. Consequently, they emphasised the necessity for professional development that not only improves teachers’ linguistic abilities but also addresses pedagogical competencies for effectively implementing EMI in their classrooms.

Participants noted the scarcity of professional development training designed to enhance teachers’ linguistic skills for facilitating the integration of EMI, particularly within CEIL. Kenza remarked, “I think there are a few initiatives which aim to provide language support”. This comment underscores the limited scope of current initiatives, which fail to sufficiently address the broader needs of teachers required to deliver content in English. Several interviewees expressed concerns regarding the efficacy of these training sessions, pointing out that they are predominantly aimed at enhancing teachers’ general English skills rather than focusing on ESP. Huyem highlighted this issue, asserting, “The lack of effective training of teachers and educators, especially in ESP”, which implies that without targeted training in ESP, teachers are ill-equipped to handle subject-specific language requirements.

Nour further elaborated on this inadequacy, stating, “Training in general English via the centres for intensive language teaching across all Algerian universities”. This indicates a prevalent focus on general English training that does not align with the specialised needs of different academic disciplines. Similarly, Amar noted, “[S]ome subject-matter instructors (SMIs) studied English in CEILs but I think this is not enough to train them on how to use English as a medium of instruction”. Amar’s observation points to a gap between basic language training and the more complex requirements of delivering subject content in English.

Some participants acknowledged the availability of online platforms, such as Massive Open Online Courses (MOOCs), which aim to enhance teachers' English skills. However, they also noted the absence of professional development programmes geared towards the effective delivery of their subjects in English. Nour stated, "[O]nline platforms specifically designed for enhancing teachers' language command (i.e., reading, writing, speaking and listening) but not English instruction command". This distinction highlights the gap between general language proficiency and the ability to instruct effectively in English. Hana echoed this sentiment, asserting, "There is a lack of preparation regarding delivering subjects in English". Nour added, "[T]eachers are not receiving any tailored training that is specific to their needs", citing an example where mathematics teachers are not being trained on how to facilitate their subject using EMI: "Mathematics teachers are not receiving any specialised training on how to instruct mathematics in English or how to use the language in providing content or facilitating students' learning".

The interviewees proposed several strategies for addressing these gaps, including the organisation of training sessions focused on enhancing both language and teaching skills. Nessma recommended, "Providing professional development opportunities for instructors to enhance their English language skills and teaching methodologies". This suggests a dual approach to professional development, emphasising both linguistic and pedagogical enhancement. Fatima, a student, similarly advocated for comprehensive training, stating, "Teachers ought to be prepared to effectively deliver their subjects in English by undergoing comprehensive training that should cover their language proficiency as well as pedagogical strategies for delivering learning material in English".

To further support the successful implementation of EMI, Amina suggested forming partnerships with foreign universities and offering scholarships to educators. She stated, "Partnerships with foreign universities, offering scholarships for educators to enhance their English proficiency and teaching methodologies through specialised training programmes". Amel proposed leveraging foreign expertise, suggesting, "[F]oreign expertise must be used". Kenza highlighted the importance of training in using technological aids, recommending, "[R]eceiving training in the use of technology tools and digital resources to enhance English-medium instruction. This includes training on interactive whiteboards, educational software, online learning platforms, and multimedia resources". Souheyla and Farah emphasised the importance of collaborative support among lecturers. Souheyla recommended, "collaboration between colleagues", while Farah echoed this, advocating for the creation of support networks and development programmes: "[C]reating professional development programs, workshops, and support networks can help educators enhance their language proficiency and teaching skills".

The findings from the interviews clearly indicate a significant deficiency in quality professional development for teachers aimed at EMI. Current training programmes tend to focus narrowly on language enhancement, with little to no emphasis on the skills required to teach content through EMI effectively. While improving staff language proficiency is vital, it must be coupled with the enhancement of their teaching methodologies to ensure effective EMI implementation. Strategies such as organising workshops, seminars, providing scholarships, and fostering collaborations with international institutions were suggested as effective means to enable teachers to successfully deliver their subjects through English. Ensuring equal access to affordable and high-quality training is crucial for SD, aligning with the objectives of SDG 4.3, which aims to provide equal access to affordable and quality education.

5.6 The Future of Implementing EMI in Algeria

After examining the current state of EMI in Algerian HEIs, it becomes imperative to delve into the perspectives of teachers and students regarding the future integration of EMI. This exploration is crucial within the context of this book, which aims to scrutinise the concept of sustainability in Algerian higher education. Specifically, it seeks to investigate the balance between recent reforms focused on digitalisation (as detailed in Chap. 4) and the implementation of EMI (explored in this chapter), along with their associated practices and outcomes. Sustainability, as a foundational concept, entails the responsible management of resources over time across economic, social, and environmental dimensions. The UN Brundtland Commission's seminal definition in 1987 frames SD as meeting current demands while ensuring future generations can fulfil their own requirements. This chapter concentrates on investigating whether the adoption of EMI can serve as an indicator of progress towards SDG4.

SDG4 underscores the commitment to promoting accessible and fair-quality education while advocating for lifelong learning opportunities for all. Remarkably, previous studies have not thoroughly addressed whether the implementation of EMI aligns with SDG4 objectives, making this investigation particularly novel and pertinent. Through data analysis, two principal themes emerged concerning the future of EMI in Algeria: educational and vocational prospects, and EMI's role as a pathway to GC. These themes will be meticulously examined in subsequent subsections, drawing upon the voices of participants to elucidate their perspectives.

In examining educational and vocational prospects, the focus will be on how EMI can enhance students' readiness for global employment markets and contribute to the development of a skilled workforce aligned with national and international demands. Additionally, exploring EMI as a gateway to GC entails investigating how proficiency in English can foster intercultural understanding, global awareness, and participation in international academic and professional communities. Aligning these themes with SDG4 necessitates an exploration of the relevant targets and strategies for achieving them. This involves scrutinising how EMI implementation can support equitable access to quality education, enhance educational outcomes, and foster lifelong learning opportunities that are inclusive of diverse learner needs.

By foregrounding the voices of teachers and students, this section aims to provide a nuanced understanding of the future of implementing EMI in Algerian HEIs. It seeks to contribute to ongoing debates about the role of language policies in educational reform and their implications for SDGs. Through this exploration, we aspire to illuminate pathways towards a more inclusive, globally connected HE landscape in Algeria, grounded in principles of sustainability and educational equity.

5.6.1 *Educational and Vocational Prospects*

The integration of EMI within Algerian HEIs faces significant challenges. Participants in this study have identified several obstacles, including insufficient resources, linguistic marginalisation, and inadequate professional development for educators, which impede the effective implementation of EMI. Despite these challenges, there is a sense of optimism among some participants about the potential for EMI's growth, provided these issues are addressed effectively.

One of the participants, Nezha, posits, "I think that the future of English-medium instruction in Algerian higher education is likely to see continued growth, but with a focus on addressing current challenges and ensuring its effectiveness". Nezha underscores that overcoming the present challenges is crucial for realising effective EMI practices and outcomes. Huyem echoes this sentiment by stating, "I think EMI in Algeria is a very promising project. Trainings for educators are being held (although they need to be improved), extra credit for teaching/publishing in English across different fields, etc." Huyem suggests that successful EMI integration relies on providing qualified educators, enhanced through improved training programmes, and incentivising teaching and publishing in English. This perspective aligns with the third means of implementing SDG4, specifically target 4.c, which calls for effective training to prepare qualified teachers.

Abderrezzaq highlights another critical aspect by stating, "EMI has a promising future in Algeria only if the environment proves to be supportive". This comment reflects target 4.a of SDG4, which emphasises the need for inclusive and effective learning environments. Yasmine, a student, shares a similar perspective: "[T]he future of EMI in Algeria appears promising to me. While there are challenges to address, the potential benefits for students, faculty members, and the higher education system as a whole are significant and worthwhile". This optimism suggests that, while challenges exist, the benefits of integrating EMI such as improved educational quality and greater opportunities for students and faculty are significant and compelling.

Several participants believe that EMI could broaden educational opportunities for students. Nour articulates, "I think that the implementation of EMI can allow students to access resources and content at the international level, enhance their 21st-century skills, and participate in the globalised world". This suggests that EMI can provide students with greater access to global resources and prepare them with skills essential for success in a globalised context. Nezha proposes that setting classroom guidelines based on mutual respect could foster a safe learning environment: "[E]stablishing classroom norms that promote respect, valuing diverse perspectives, and using kind language helps students feel comfortable participating and expressing themselves, regardless of their English proficiency". By promoting mutual respect and valuing diversity, EMI classes can become spaces that support both teachers and students, contributing to social sustainability.

Fatima supports this view, stating, "[T]eachers should foster a classroom culture that values diversity and promotes mutual respect among students". Ahlem suggests that using diverse materials and multiple languages to simplify complex concepts could create a more inclusive atmosphere in EMI classes: "[T]he use of various resources using different languages. Simplifying concepts using simple language". This reflects the need to balance EMI's content delivery with the diverse linguistic backgrounds of students. Nihed highlights the challenge of integrating Algeria's linguistic diversity into EMI practices: "EMI policies often adhere to monolingual ideologies, assuming that English can be the sole medium of instruction; however, Algerian students have varying language backgrounds, so balancing these ideologies with practical implementation can be challenging". This underscores the difficulty of applying a strict English-only policy in a multilingual context like Algeria.

Ryad adds to this perspective, "I think teachers themselves are at a loss and are too looking for safe learning environments because they have diverse language backgrounds". This suggests that teachers also face challenges in navigating the multilingual realities of their classrooms. Amina advocates for employing students' diverse linguistic backgrounds and using visual aids and bilingual resources to make content accessible to all students, regardless of their English proficiency: "[C]ultivate a supportive classroom culture where multilingualism is valued and celebrated". This perspective indicates that a multilingual classroom culture is fundamental to the success of EMI. Wafa supports this by sharing her classroom practice: "[I]n my classroom, I encourage translanguaging as a practice to develop English and other languages; after all, literacy is what improves language". These practices reflect target 4.a of SDG4, which advocates for educational facilities and safe, inclusive learning environments. The strategies identified by participants including mutual respect, diverse resources, and tailored feedback are seen as crucial for enhancing educational prospects and sustaining EMI integration in Algerian HEIs.

Participants also argue that these strategies can promote equitable and inclusive access to educational and vocational opportunities. Amina contends, "[E]nsuring students from diverse linguistic and socio-economic backgrounds can develop strong English language skills, enabling them to pursue vocational training and higher education both domestically and internationally". This perspective suggests that EMI can play a crucial role in levelling the educational playing field for students from various backgrounds. Nessma adds, "[T]he effective integration of EMI ensures equitable access to vocational and educational opportunities". She believes that EMI classes can improve students' English skills, thereby providing them with broader access to educational and vocational resources and technologies. This view aligns with target 4.3 of SDG4, which promotes equal access to affordable technical, vocational, and higher education. Huyem further asserts, "I think being taught in English would enhance the international employability of Algerian students and help them integrate easily into different global workplaces/opportunities". This highlights the role of EMI in increasing students' chances of accessing the global job market, aligning with target 4.4 of SDG4, which focuses on increasing the number of individuals with skills relevant for employment, thus contributing to economic sustainability.

The integration of EMI in Algerian HEIs presents both significant opportunities and formidable challenges. While there is a strong belief in EMI's potential to enhance educational and vocational prospects, its successful implementation requires addressing key issues such as resource limitations, linguistic diversity, and the necessity for robust professional development. Creating inclusive and supportive learning environments and employing diverse pedagogical strategies are crucial for the effective integration of EMI. Incorporating a bottom-up approach in policymaking, one that includes the experiences and voices of educators and students, can help reconcile reforms with practical classroom realities. This approach is essential for ensuring a more resilient and effective integration of EMI, aligning with the targets of SDG4, which aim to provide inclusive, equitable, and quality education while promoting lifelong learning opportunities for all.

By addressing the challenges and leveraging the diverse linguistic and cultural strengths of Algerian students, EMI can play a pivotal role in preparing students for success in an increasingly interconnected and globalised world. The study by Pun et al. (2022) questions the sustainability of top-down EMI policies in science classrooms at a Hong Kong secondary school. Drawing from these insights, we emphasise the importance of incorporating teachers' and students' perspectives to balance reforms and achieve practical, effective outcomes in Algerian HEIs. This approach underscores the significant benefits of EMI integration policies that consider and incorporate the practical experiences and needs of those directly involved in the educational process.

5.6.2 EMI as a Gateway to Global Citizenship

The interviewees asserted that the effective implementation of EMI in Algerian HEIs enables students to communicate in intercultural encounters and engage with global issues, thus evolving into global citizens. Baker and Fang (2021) identified that participating in EMI programmes at international universities abroad is linked to GC. However, they emphasised that the development of GC through EMI experiences is not guaranteed, suggesting a need for formal education in GC. Salem (2021) echoed this recommendation, observing that Algerian EFL teachers and students demonstrated a limited understanding of global citizenship education (GCE) due to minimal exposure, advocating for its inclusion in academic curricula.

Several interviewees highlighted that EMI classes have the potential to prepare students for GC by enhancing their language skills, intercultural communicative competence, and international perspectives. Given that English is extensively used globally, they noted that EMI classes could enable students to participate effectively in global contexts. Fatima stated, "[B]y being proficient in English, which is widely recognised as the global language of communication, students can enhance their ability to interact with people from diverse linguistic backgrounds, collaborate on international projects, and participate in global forums". She emphasised that EMI programmes which expose students to global issues can foster GC. Fatima further remarked, "[S]tudents gain insights into different cultures, customs, and ways of thinking. This exposure promotes cultural awareness, empathy, and appreciation for diversity, enabling students to navigate multicultural environments and interact respectfully with individuals from around the world". She argued that EMI classes have the potential to nurture empathy and respect for diversity, essential for co-existing in a multicultural world.

Similarly, Kenza observed that EMI programmes could enhance global awareness and understanding, allowing students to navigate cultural differences and engage respectfully with individuals from different cultural backgrounds in discussions and projects on global issues. This, she asserted, encourages them to become responsible global citizens who contribute to the well-being of society, which is crucial for social sustainability. Kenza articulated, "EMI develops students' English language proficiency, enabling them to communicate effectively with people from diverse linguistic backgrounds around the world. EMI exposes students to diverse cultural perspectives, values, and norms, fostering cultural awareness and appreciation". Nessma shared this perspective, suggesting that EMI could help students engage with global concerns and contribute positively to the world. She contended that EMI classes have the potential to prepare students for GC by "equipping them with English language proficiency, access to international resources and perspectives, and opportunities for cross-cultural interaction and collaboration".

Abderrezzaq also mentioned that EMI classes could foster students' GC by enhancing their cultural awareness: "[A]cquiring a new language makes students more culturally aware. They become the citizens of the world". Huyem expressed a similar view, considering EMI classes as opportunities to develop students' intercultural communication skills, enabling them to access various academic resources within the global academic landscape. She noted, "EMI will also help them foster intercultural competence and communication skills (being able to chase scholarships and participate in international events and conferences)". Farah suggested that students learning their disciplines in English would learn about the cultural aspects of the language, thus helping them become global citizens: "[B]ecause they learn the English language, they will be exposed to its culture. They will use English to communicate effectively with people/countries all over the world". Naji emphasised that through EMI programmes, students could work in diverse settings worldwide: "EMI produces universal students. They can work anywhere in the world". Anes, a student, asserted that language is essential for students to grow as global citizens: "[L]anguage is the basis of the student's self-development". Yasmine, another student, believed that EMI classes could enhance Algerian students' language skills and interculturality, enabling them to interact with people from different cultures. She remarked, "EMI can help improve Algerian students' proficiency in English which, in turn, will allow them to communicate and collaborate with people from diverse backgrounds on a global scale and develop their intercultural competence".

However, Nour noted that the current implementation of EMI cannot truly prepare students for GC due to existing barriers. She stated, "[S]o far, I do not see how it is fostering Algerian students' abilities given that its application is facing a lot of challenges and hindrances". Nour suggested that addressing these obstacles could facilitate the effective integration of EMI, helping students to engage with the world and utilise available resources: "[I]f measures are taken to improve the situation, I think EMI might be beneficial in opening the doors for Algerian university students to integrate with the global world and access content that is available out there". This underscores the necessity of considering the voices of teachers and students to enhance the implementation of EMI in Algerian HEIs.

These findings suggest that EMI can act as a conduit for developing GC by improving students' communication skills, intercultural competence, and awareness of global issues. Participants recognised that language and culture are intertwined, and EMI cannot be implemented effectively without considering the cultural dimensions of language, an idea extensively discussed by leading scholars in the fields of language and culture (e.g. Byram, 1997; Kramsch, 1998). The results indicate that learners can acquire the necessary knowledge, skills, and values for GC through EMI programmes, aligning with target 4.7 of SDG4, provided the current obstacles are addressed. Achieving this requires adequate resources, creating safe and inclusive multilingual learning environments, and offering qualified teachers through affordable trainings and equitable access to scholarships, all of which align with the objectives of SDG4, targets 4.a, 4.b, and 4.c.

5.7 Conclusion

This chapter presented the findings from interviews conducted with 31 participants, comprising 23 teachers and 8 students, from various Algerian universities. The focus of these interviews was on the integration of EMI within Algerian HEIs. In Algeria, Arabic and Tamazight are the official languages, while French is also widely spoken due to the historical context of colonisation.

This multilingual landscape is mirrored within the educational system: French serves as the medium of instruction for scientific disciplines, whereas Arabic is predominantly used for the social sciences and humanities. English, though minimally utilised in daily societal interactions, is taught as a subject in middle and secondary education and is incorporated as both a subject within numerous university disciplines and a standalone field of study.

Following the Al Hirak movement of 2019, there emerged a significant interest in fostering a generation proficient in English. Consequently, English was introduced into the curriculum of primary schools in 2022 and began to be employed as a medium of instruction in higher education, aimed at increasing internationalisation and enhancing the visibility of research. Additionally, the adoption of EMI appears to be a strategy to decolonise education by diminishing the centrality of French. To facilitate this transition, training programmes were organised for teachers to equip them with the skills to deliver their course content in English. Despite Algeria's inherent multilingualism, the current trend seems to favour a monolingual approach, advocating for an English-only policy within higher education. At present, university educators are encouraged to teach and publish their research in English. Students who compose their graduation dissertations in English are celebrated and their works are widely shared on various social media platforms.

The incorporation of English into the Algerian educational system has progressed from teaching EFL to encompassing ESP. More recently, EMI has been introduced, prompting debates regarding the efficacy of CLIL in improving both subject-based learning and English proficiency within HEIs. Exploring the implementation of EMI in Algerian HEIs is essential for discerning its long-term viability, particularly in the context of the UN's SDG4 and its associated targets. This area remains significantly under-researched, highlighting the need for comprehensive studies to examine its implications for educational practices and outcomes. The thematic analysis of interview data identified two primary themes, each comprising several sub-themes: the present situation of EMI adoption in Algerian HEIs and the prospective future of EMI implementation in Algeria. Within the first theme, our exploration focused on exploring the alignment of current EMI practices with existing demands. Meanwhile, the second theme delved into examining the potential of EMI to address the future needs of upcoming generations. These themes align closely with the UN's definition of sustainability, emphasising the balance between meeting immediate requirements and ensuring that future generations can fulfil their own needs. In this book, sustainability is considered to involve social, economic, and environmental dimensions, as previously mentioned.

Although the expansion of English as a global lingua franca partly stems from colonialism, the introduction of EMI in Algeria is viewed by participants as a strategic effort to "defrenchify" the nation's educational and research systems, moving away from their colonial Francophone heritage. This policy aims to decolonise education and is seen as a means to enhance international opportunities and employment prospects. The findings indicate that EMI's adoption is intended to improve the quality of education, ensuring accessibility and inclusivity for international students. This aligns with the principles of SDG4, which promote inclusive and equitable quality education and lifelong learning opportunities for all. By implementing EMI, Algeria aspires to increase the global visibility of its academic research and provide its students with greater access to international experiences, including the global job market. This approach not only supports social, economic, and environmental sustainability but also strengthens Algeria's engagement with the international academic and professional communities.

However, the participants observed that the current status of the integration of EMI in Algeria remains nascent. The shift to English, following the long-standing usage of Arabic and French, presents significant challenges for both educators and students. Participants proposed strategies such as translanguaging, the utilisation of diverse educational materials, and the adoption of ESP and CLIL to facilitate the effective use of EMI and the comprehension of content. They highlighted linguistic marginalisation, a scarcity of resources, and the insufficient professional development of teachers as substantial impediments to the successful implementation of EMI in Algeria. These barriers underscore a broader issue within the educational system, where a lack of alignment between policy and practice exacerbates existing inequities. The findings indicate a shortfall in the practical application of SDG4 targets (4.a, 4.b, and 4.c), which could adversely impact educational outcomes. This discrepancy between policy intentions and practical realities calls into question the efficacy of current efforts to meet international educational standards. Although not formally articulated, the policy of introducing EMI in Algeria ostensibly supports SDG4. Nevertheless, the current landscape suggests a misalignment between practice and the principles of SDG4, with the reform failing to adequately address the needs of the contemporary generation.

The participants endorsed the integration of EMI in Algeria, asserting that its future implementation could fulfil the educational and vocational needs of subsequent generations. They believed that addressing the current obstacles associated with EMI could significantly enhance both educational and career prospects, positioning it as a gateway to GC. This aligns with the broader objectives of SDG4, suggesting that the strategic application of EMI could substantially contribute to SD within the country. However, the participants also raised concerns regarding the hasty introduction of EMI without sufficient preparatory measures. They cautioned that such rapid adoption might disadvantage teachers and students who are not proficient in English. To mitigate this risk, there is an urgent imperative to create inclusive and safe learning environments that respect and incorporate the diverse linguistic backgrounds of all students, as outlined in SDG 4.a. Expanding access to scholarships (SDG 4.b) and significantly increasing the supply of qualified teachers through robust and comprehensive teacher training programmes (SDG 4.c) are critical steps to ensure that EMI can be implemented effectively and equitably. It is essential to recognise that the integration of EMI, while aligned with global educational aspirations, necessitates meticulous planning, adequate resource allocation, and a commitment to inclusivity.

The incorporation of EMI into Algerian HEIs necessitates strategic planning and rigorous implementation. Such an approach promises not only to enrich the educational landscape but also to contribute significantly to broader social, economic, and environmental sustainability. Successful adoption of EMI holds the potential to foster a more interconnected and globally conscious society, contingent upon meeting the diverse needs of students and educators through targeted and inclusive educational policies. The findings outlined in this chapter indicate that adopting a collaborative strategy, integrating perspectives from both grassroots (teachers and students) and top-level (policymakers and decision-makers) stakeholders in policy formulation, is crucial for achieving a balanced approach to educational reforms, practices, and outcomes.

References

Airey, J. (2016). EAP, EMI or CLIL? In K. Hyland & P. Shaw (Eds.), *The Routledge handbook of English for academic purposes* (pp. 71–83). Routledge.

Alhassan, A. (2021). Challenges and professional development needs of EMI lecturers in Omani higher education. *Sage Open, 11*(4), 21582440211061527.

Alkhateeb, H., & Bouherar, S. (2022). Sustainability and educational language policy in Arab higher education: Findings from Q research. *Current Issues in Language Planning, 23*(5), 1–23.

Baker, W., & Fang, F. (2021). 'So maybe I'm a global citizen': Developing intercultural citizenship in English medium education. *Language, Culture and Curriculum, 34*(1), 1–17.

Bälter, O., Kann, V., Mutimukwe, C., & Malmström, H. (2023). English-medium instruction and impact on academic performance: A randomized control study. *Applied Linguistics*. https://doi.org/10.1515/applirev-2022-0093

Banegas, D. L., & Busleimán, G. M. (2021). EMI materials in online initial English language teacher education. In D. Lasagabaster & A. Doiz (Eds.), *Language use in English-medium instruction at university: International perspectives on teacher practice* (pp. 100–125). Routledge.

Belmihoub, K. (2018). English in a multilingual Algeria. *World Englishes, 37*(2), 207–227.

Bennoune, M. (2000). *Education, culture et droeloppement en Algerie: Bilan et perspectives du systeme educatif.* Marinoor-ENAG.

Benrabah, M. (2005). Language planning situation in Algeria. *Current Issues in Language Planning, 6*(4), 379–502.

Benrabah, M. (2013). *Language conflict in Algeria: From colonialism to post-independence*. Multilingual Matters.

Bentahar, Z. (2020). "Ytnahaw ga'!": Algeria's cultural revolution and the role of language in the early stages of the spring 2019 Hirak. *Journal of African Cultural Studies, 33*(4), 471–488.

Bouherar, S., & Ghafsi, A. (2021). *Algerian languages in education: Conflicts and reconciliation*. Palgrave Macmillan.

Branka, D. M., & Molino, A. (2022). Translanguaging in EMI: Lecturers' perspectives and practices. *Ricognizioni, 9*(17), 29–47.

Braun, V., & Clarke, V. (2006). Using thematic analysis in psychology. *Qualitative Research in Psychology, 3*(2), 77–101.

Brinton, D., Snow, M. A., & Wesche, M. B. (2003). *Content-based second language instruction*. University of Michigan Press.

Bruton, A. (2011). Is CLIL so beneficial, or just selective? Re-evaluating some of the research. *System, 39*(4), 523–532.

Bruton, A. (2013). CLIL: Some of the reasons why... and why not. *System, 41*(3), 587–597.

Bryman, A. (2016). *Social research methods* (5th ed.). Oxford University Press.

Byram, M. (1997). *Teaching and assessing intercultural communicative competence*. Multilingual Matters LTD.

Canagarajah, S. (2011). Codemeshing in academic writing: Identifying teachable strategies of translanguaging. *The Modern Language Journal, 95*(3), 401–417.

Carrió-Pastor, M. L. (2021). CLIL vs EMI: Different approaches or the same dog with a different collar? In M. L. Carrió-Pastor & B. Bellés Fortuño (Eds.), *Teaching language and content in multicultural and multilingual classrooms: CLIL and EMI approaches* (pp. 13–30). Palgrave Macmillan.

Costa, F., & Mastellotto, L. (2022). The role of English for specific purposes (ESP) in supporting the linguistic dimension in English-medium instruction (EMI). *CLIL: Journal of Innovation and Research in Plurilingual and Pluricultural Education, 5*(2), 37–52.

Coyle, D., Hood, P., & Marsh, D. (2010). *Content and language integrated learning*. Cambridge University Press.

Dafouz, E. (2018). English-medium instruction in multilingual university settings – An opportunity for developing language awareness. In P. Garrett & J. M. Cots (Eds.), *The Routledge handbook of language awareness* (1st ed., pp. 170–185). Routledge.

Dafouz, E. (2021). Crossing disciplinary boundaries: English-medium education (EME) meets English for Specific Purposes (ESP). *Ibérica, 41*, 13–38.

Daoudi, A. (2018). Multilingualism in Algeria: Between 'soft power', 'Arabisation', 'Islamisation', and 'globalisation'. *The Journal of North African Studies, 23*(3), 460–481.

Dearden, J. (2014). *English as a medium of instruction-a growing global phenomenon*. British Council. https://www.britishcouncil.es/sites/default/files/british_council_english_as_a_medium_of_instruction.pdf

Denscombe, M. (2007). *The good research guide for small-scale social research projects* (3rd ed.). Open University Press.

Diener, E., & Crandall, R. (1978). *Ethics in social and behavioral research*. the University of Chicago Press.

Etikan, I., Musa, S. A., & Alkassim, R. S. (2016). Comparison of convenience sampling and purposive sampling. *American Journal of Theoretical and Applied Statistics, 5*(1), 1–4.

Gopinathan, S. (2007). Globalisation, the Singapore developmental state and education policy: A thesis revisited. *Globalisation, Societies and Education, 5*(1), 53–70.

Guba, E. G. (1981). Criteria for assessing the trustworthiness of naturalistic inquiries. *Ectj, 29*(2), 75–91.

Gumperz, J. J. (1964). Linguistic and social interaction in two communities. *American Anthropologist, 66*(6), 137–153.

Hutchinson, T., & Waters, A. (1987). *English for specific purposes: A learning-centered approach*. Cambridge University Press.

Hüttner, J., & Smit, U. (2014). CLIL (Content and Language Integrated Learning): The bigger picture. A response to: A. Bruton. 2013. CLIL: Some of the reasons why… and why not. System 41 (2013): 587–597. *System, 44*, 160–167.

Im, J. H., & Kim, J. (2015). Use of blended learning for effective implementation of English-medium instruction in a non-English higher education context. *International Education Studies, 8*(11), 1–15.

Jacob, C. (2020). English as a decolonial language: Academic frames, popular discourses & language practices in Algeria. *The Journal of North African Studies, 25*(6), 1013–1032.

Kadri, A. (2014). Les conditions d'imposition du système scolaire français à l'Algérie. In A. Kadri (Ed.), *Instituteurs et enseignants en Algérie 1945–1975: Histoire et mémoires* (pp. 19–53). Karthala.

Kaur, J. (2020). Using English for interaction in the EMI classroom: Experiences and challenges in a Malaysian public university. In H. Bowles & A. C. Murphy (Eds.), *English-medium instruction and the internationalization of universities* (pp. 129–154). Palgrave Macmillan.

Kormos, J. (2017). *The second language learning processes of students with specific learning difficulties.* Routledge.

Kramsch, C. (1998). *Language and culture.* Oxford University Press.

Kumar, R. (2011). *Research methodology: A step-by-step guide for beginners* (3rd ed.). Sage.

Le Hoang Ngo, P. (2021). Translanguaging practices in EMI settings from the perspective of student agency: An example from Vietnamese higher education. In W. Tsou & W. Baker (Eds.), *English-medium instruction translanguaging practices in Asia: Theories, frameworks and implementation in higher education* (pp. 121–141). Springer.

Le Roux, C. S. (2017). Language in education in Algeria: A historical vignette of a 'most severe' sociolinguistic problem. *Language & History, 60*(2), 112–128.

Lin, T., & Lei, J. (2021). English-medium instruction and content learning in higher education: Effects of medium of instruction, English proficiency, and academic ability. *Sage Open, 11*(4) https://journals.sagepub.com/doi/pdf/10.1177/21582440211061533

Macaro, E. (2018). *English medium instruction.* Oxford University Press.

Mackey, W. (1989). Determining the status and function of languages in multinational societies. In U. Ammon (Ed.), *Status and function of languages and language varieties* (pp. 3–20). De Gruyter.

Maraf, B., & Vanci Osam, U. (2022). The smile revolution (hirak) as a driving force for an English 'tidal wave' and foreign language policy-making in Algeria. *Current Issues in Language Planning, 24*(2), 179–200.

Marginson, S. (1999). After globalization: Emerging politics of education. *Journal of Education Policy, 14*(1), 19–31.

Marginson, S., & Rhoades, G. (2002). Beyond national states, markets, and systems of higher education: A glonacal agency heuristic. *Higher Education, 43*, 281–309.

O'Neill, M., & Chapman, A. (2015). Globalisation, internationalisation and English language: Studies of education in Singapore, Malaysia and Australia. *Education Research and Perspectives, 42*, 1–24.

Phuong, Y. H., & Nguyen, T. T. (2019). Students' perceptions towards the benefits and drawbacks of EMI classes. *English Language Teaching, 12*(5), 88–100.

Pun, J., Thomas, N., & Bowen, N. E. J. A. (2022). Questioning the sustainability of English-medium instruction policy in science classrooms: Teachers' and students' experiences at a Hong Kong secondary school. *Sustainability, 14*(4), 2168.

Räisänen, C., & Fortanet-Gómez, I. (2008). The state of ESP teaching and learning in Western European higher education after Bologna. In C. Räisänen & I. Fortanet-Gómez (Eds.), *ESP in European higher education: Integrating language and content* (pp. 11–51). John Benjamins B.V.

Rose, H., Curle, S., Aizawa, I., & Thompson, G. (2019). What drives success in English medium taught courses? The interplay between language proficiency, academic skills, and motivation. *Studies in Higher Education, 45*(11), 1–14.

Salem, S. (2021). *Global citizenship education in English as a foreign language (EFL) classrooms: Views of the Algerian university oral expression teachers and their students.* Doctoral dissertation, University of York.

Siegel, J. (2020). Comprehension in English Medium Instruction (EMI) lectures: On the impact of lecturer L2 English use. *Language Learning in Higher Education, 10*(1), 73–91.

Smit, U. (2019). Classroom discourse in EMI: On the dynamics of multilingual practices. In K. Murata (Ed.), *English-medium instruction from an English as a lingua franca perspective* (pp. 99–122). Routledge.

Snow, M. A. (2001). Content-based and immersion models for second and foreign language teaching. In M. Celce-Murica (Ed.), *Teaching English as a second or foreign language* (3rd ed., pp. 303–318). Heinle & Heinle Thompson Learning.

Stake, R. E. (1978). The case study method in social inquiry. *Educational Researcher, 7*(2), 5–8.

The New Arab. (2022). *Algeria recruits 5,000 new English language teachers for primary schools in shift from French.* https://www.newarab.com/news/algeria-hires-5000-english-teachers-shift-french.

Tilmatine, M. (2015). Arabisation and linguistic domination: Berber and Arabic in the North of Africa. In C. Stolz (Ed.), *Language empires in comparative perspective* (pp. 1–16). De Gruyter.

Tran, N. T., Tran, T. B. T., & Bien, T. T. M. (2020). An exploration of the factors hindering students' lesson comprehension in EMI classes. *International Journal of Language and Literary Studies, 2*(3), 29–42.

Van der Walt, C. (2013). *Multilingual higher education: Beyond English medium orientations* (Vol. 91). Multilingual Matters.

Vu, N. T., & Burns, A. (2014). English as a medium of instruction: Challenges for Vietnamese tertiary lecturers. *Journal of Asia TEFL, 11*(3), 1–31.

Weber, J. J., & Horner, K. (2012). *Introducing multilingualism: A social approach.* Routledge.

Yilmaz, T., Li, S., & Lee, Y. J. (2024). A multilingualism-as-resource approach in a graduate-level English-medium instruction (EMI) program at a Korean university. *Journal of Multilingual and Multicultural Development, 45*(1), 1–14.

Yin, R. K. (2003). *Case study research: Design and methods* (3rd ed.). SAGE Publications.

CHAPTER 6

Sustain(able) Higher Education Practices

6.1 Summary of the Key Findings

The evolution of sustainability as both a term and a practical concept has been extensively analysed and discussed within the literature. Within the realm of education, sustainability has branched into various directions, culminating in a call for interdisciplinary approaches where problem-based learning serves both as a goal and a method (Dale & Newman, 2005; Moore, 2005). Scoffham and Rawlinson (2022) introduce the term "hybrid concept" to encompass the diverse dimensions that define sustainability. In our study, we have elucidated sustainability and our approach to examining its mechanisms in the context of Algerian higher education (HE). Our understanding of sustainability, as articulated in the Brundtland Report of 1987, revolves around meeting current socio-economic needs without compromising resources or causing environmental harm that would hinder future generations from meeting their own needs. It emphasises responsible, long-term planning to ensure the well-being of both present and future generations. As such, the effectiveness of this definition is benchmarked against the United Nations' Goal 4 of quality education. This conceptual framework forms the heart of our subsequent discussions, serving as the foundational principle upon which all further analysis and discourse are based. Through this lens, we aim to explore how sustainability can be effectively integrated into HE practices in Algeria, ensuring that educational systems not only meet immediate socio-economic demands but also contribute to a sustainable future.

Measuring this framework against previous HE reforms reveals a state of confusion regarding the independence of universities due to their strong ties to politics (Bouherar & Ghafsi, 2021). Rather than fostering sustainable development (SD) and integrating the principles of sustainability, these reforms were characterised by periods of detachment defined by clashing elites and their control over the country's ruling mechanisms. The cooperation between

S. Bouherar, S. Salem, *Sustainability in Algerian Higher Education*, https://doi.org/10.1007/978-3-031-81347-4_6

France and Algeria in education (Matasci et al., 2020) marked a significant era in the history of Algerian education, paving the way for other unstable reforms in the 1970s. French attempts to gallicise the Algerian educational system (Khandriche, 1996) succeeded in undermining the credibility of subsequent reforms. The reforms of the 1970s, 1980s, and 2000s reflected socio-political changes but lacked proper efforts to work towards a structured plan for long-term sustainability.

Sustainable Development Goal 4 (SDG4) was also examined against the aims of digitalisation in HE, revealing significant insights. Primarily, the digitalisation efforts in HE were more closely aligned with administrative functions than with the academic domain. This misalignment resulted in substantial challenges for pedagogical practices. E-learning initiatives, in particular, suffered considerably due to inadequate technological infrastructure and insufficient investment in faculty development and training programmes (such as scholarships and partnerships with other international universities) designed for e-learning purposes. The lack of a robust digital infrastructure meant that many institutions struggled to implement effective e-learning solutions, thereby limiting the accessibility and quality of education delivered through digital platforms. Furthermore, the insufficient investment in training faculty members to adapt to and effectively utilise digital tools hindered the potential benefits of e-learning, not to mention the lack of technological equity and inclusion among students (Grimes & Porter, 2023). Faculty members often faced challenges in integrating digital technologies into their teaching methodologies, resulting in suboptimal educational experiences for students and often leading them to resort to traditional teaching methods. As a result, the aspirations for digitalisation within HE have not yet facilitated a smooth transition to quality education. This shortfall undermines the broader goals of SD, as quality education is a critical component of achieving long-term sustainability outcomes. The current state of digitalisation, therefore, highlights the need for a more comprehensive and well-funded approach to developing digital infrastructure and training programmes that prioritise academic excellence alongside administrative efficiency.

Another element we have benchmarked against quality education is the implementation of English as a medium of instruction (EMI), which appeared to have decolonial purposes (Jacob, 2020) against the French language. The swift transition to English has resulted in a dominant monolingual approach in HE, where the use of English has been unconditionally advocated over other languages. This abrupt total inclusion of English has led to linguistic marginalisation of speakers of other languages in HE. Furthermore, the current linguistic landscape in HE indicates a disconnect between actual practices and the principles of sustainable development goal 4 (SDG4). The recent reform efforts have not sufficiently addressed the needs of the modern student population. The abrupt implementation of EMI, without adequate preparation, has led to considerable issues. The swift adoption of English may place teachers and students who lack proficiency at a disadvantage, creating obstacles to effective teaching and learning. To achieve Education for Sustainable Development (ESD) reforms, a more balanced approach is essential. Reforms can be made more inclusive, taking into account linguistic diversity and the educational needs of all students.

On a broader scale, the discourse on sustainability implementation in Algerian HE is still in its early stages due to partial implementations and deficiencies in practice. The question posed earlier in the introductory chapter about whether Algerian HE adheres to the principles of SDG4 quality education can be addressed to some extent. While sustainable intentions are present, the implementations may not consistently reflect a diligent approach towards sustainability. One significant issue is the lack of a cohesive strategy that integrates sustainability into all aspects of HE, from curriculum design to administrative policies. The existing efforts often appear fragmented and reactive, rather than proactive and comprehensive. This approach results in temporary measures that do not significantly contribute to long-term sustainability efforts. For instance, while some institutions may introduce digital initiatives or sustainability courses, these are not always part of a broader, institution-wide commitment to sustainability, given the severe lack of technological infrastructure and digitalisation equity and inclusion. Moreover, the reliance on short-term solutions can undermine the ESD project. Temporary palliatives, such as superficial and limited e-learning platforms, fail to address the underlying issues that hinder SD. Without sustained investment in infrastructure, quality faculty training, and student engagement, sustainability efforts are unlikely to produce meaningful and lasting change. The challenges are further compounded by the economic context in Algeria, which can influence the consistency and effectiveness of sustainability initiatives. While improved economic growth can contribute to attain SDG4 (Piao & Managi, 2023), economic constraints can often lead to shifts in priorities and resources, disrupting long-term planning and implementation. This environment makes it difficult to establish and maintain the steady progress needed for true sustainability in HE.

6.2 Recommendations for Future Research/Action

Given that this book aims to explore the extent to which SDG4 quality education has been implemented in Algerian HE, it is important to highlight certain recommendations in this chapter. These recommendations are derived from the nature of the study, the identified shortcomings, and the results obtained from the research. One recommendation for future research would suggest a more detailed and specific study that could provide greater insight into how SDG4 principles are, or can be, implemented by analysing the curriculum sections that focus on teaching sustainability and assessing their credibility and practice. This involves examining whether the curriculum effectively integrates inclusivity, equity, and quality. The current study takes a broader perspective, encompassing educational reforms and newly implemented policies. However, policies alone are insufficient to comprehensively explore the SDG4 targets and means of implementation to draw specific conclusions. A curriculum-focused approach would provide deeper understanding into the practical aspects of education, revealing the actual implementation and impact of these policies on achieving the goals of SDG4. Additionally, evaluating the curriculum's content and delivery methods can uncover gaps and areas for improvement, ensuring that inclusivity, equity, and quality are not just theoretical principles but are actively practised within the educational system. This detailed analysis is crucial for making informed recommendations that can drive meaningful change in Algerian HE, aligning it more closely with the objectives of SDG4.

Another promising avenue for future research that we recommend is evaluating the contribution of SDG4 to the local economy. Given the new economic aspirations of the post-2019 government, it is crucial to examine how economic factors may or may not contribute to ESD, particularly in light of Algeria's historically modest economic growth (Haouas et al., 2024). A model-based study could illuminate the potential synergies between economic development and ESD. By analysing these interactions, researchers can assess whether economic growth policies align with and support the goals of SDG4, or if adjustments are needed to enhance their compatibility. Such an investigation would not only provide insights into the economic impacts of educational policies but also contribute to a more integrated approach to SD in Algeria. Moreover, exploring these dynamics through a model-based approach can offer predictive insights into future scenarios, helping policymakers make informed decisions that foster both economic prosperity and educational sustainability. This research direction holds promise for advancing the understanding of how education and economic strategies can mutually reinforce SDGs in Algeria.

One of the notable shortcomings of this study is the relatively small scale of participants in the survey. Although the sample size of 404 participants is representative of nearly all public universities and includes a range of responses, a larger sample size exceeding 1000 participants would have provided more robust statistical power and reliability. Despite efforts, we faced challenges in collecting a significant number of participants and encouraging broader participation. Future studies could benefit from employing alternative techniques, such as supervised data collection methods, to enhance participant recruitment and increase sample size. These methods could include targeted outreach campaigns, collaborations with educational institutions, or leveraging digital platforms to reach a wider audience. Similarly, the number of interviews conducted (31 interviews) could have been expanded to include a more diverse range of perspectives from additional universities beyond the 20 institutions covered across the country. This broader inclusion would have enriched the qualitative insights gathered and provided a more comprehensive understanding of the implementation of SDG4 in Algerian HE. Addressing these limitations in future research conducts will be crucial for achieving a more comprehensive and nuanced exploration of SDG4 implementation in the Algerian context, ensuring that findings are both robust and reflective of the diverse realities within higher education institutions (HEIs).

Recommendations for future action offer three main suggestions. One key issue is the lack of infrastructure, as highlighted in Chap. 4. The shortage of technological materials, such as personal computers for students and adequate internet coverage, has created significant barriers to inclusion and the quality of education. The current technological infrastructure is insufficient to support widespread digital learning initiatives. As a result, efforts to implement such solutions have been met with significant challenges. To bridge this gap, there must be substantial investment in both technology and faculty training. Providing the necessary tools and knowledge to both educators and students is essential for creating an environment conducive to effective digital learning. To integrate SDG4 principles effectively, decision-makers must consider the large number of students enrolled in public universities, where quality education has been compromised due to overcrowding. Addressing this challenge requires not only reducing the student-to-faculty ratio but also developing comprehensive plans to enhance infrastructure, ensuring that all students have access to quality education. While it is highly challenging to put these perspectives into practice within the current public system, one immediate step could be to encourage the development and expansion of private institutions that adhere to high standards of teaching. These institutions could serve as models for integrating SDG4 principles and offer students alternative pathways to quality education. By supporting private institutions that prioritise high-quality inclusive education, students will have the choice to engage in an educational environment that aligns with SDGs.

The second recommendation addresses the need for a multilingual approach to HE. Since the integration of EMI in HE, both educators and students have expressed concerns about being excluded from the learning process due to their lack of English proficiency. Algerian universities now host speakers of three primary languages: Arabic, French, and English. Some individuals are bilingual, while a few others are multilingual. This linguistic diversity reflects the broader cultural and societal fabric of Algeria. The issue of exclusivity has been exacerbated by the poor quality of training provided to teachers and the total absence of training for students. The sudden shift to English, often hailed as the saviour language, has worsened the linguistic landscape by neglecting Arabic and French speakers. Stakeholders must acknowledge this linguistic diversity and the unique challenges it presents in HE. To address these challenges, decision-makers need to consider a more inclusive and gradual integration of English. A multilingual approach would recognise the importance of Arabic and French alongside English, ensuring that no student or educator is left behind due to language barriers. This approach should begin with a gradual well-planned implementation of EMI, allowing students and teachers time to adapt. Initial courses could be offered in bilingual formats, with materials available in both English and either Arabic or French. Comprehensive training programmes for both teachers and students to improve English proficiency should be developed. These programmes must be mandatory and designed to build confidence and competence in using EMI. Supportive language policies should be implemented to support the use of all three languages within the educational system. This could include bilingual or trilingual resources and administrative support to foster an inclusive environment. Promoting a culture of linguistic respect and inclusion is also essential. The use of Arabic and French should be valued alongside English, and collaborative projects and initiatives that celebrate linguistic diversity should be encouraged. Effective resource allocation is crucial for supporting multilingual education. This includes hiring bilingual or multilingual staff, providing access to language learning tools, and investing in infrastructure that supports diverse language needs. By adopting a multilingual approach, Algerian HE can better integrate English while respecting and preserving the linguistic heritage of its students and educators. This strategy not only aligns with the principles of SDG4 but also ensures a more inclusive, equitable, and effective educational environment. Embracing linguistic diversity will ultimately contribute to the quality of HE and SD, allowing all students to thrive regardless of their language background.

The third recommendation raises the issue of tokenism in sustainability practices within HE. A historical review of sustainability in HE reforms since 1962 reveals that tokenism has been a persistent issue, characterised by the temporary and superficial policies enacted by successive governments. In the post 2019 era, we have identified at least two policies aimed at advancing HE that have significant implications for the implementation of SDG4. One such policy involves the launch of e-learning platforms and the promotion of digital practices. While these initiatives are commendable in theory, they do not align with the actual conditions faced by most HEIs. The gap between the rhetoric of digitalisation and the reality on the ground is stark. Educators and students have high expectations for digital practices based on the government's discourse, but these expectations remain largely unmet. The lack of adequate infrastructure, training, and support for digital learning means that the proclaimed shift towards e-learning is more symbolic than substantive. Another example of tokenistic policy is the "English tidal wave" (Maraf & Vanci Osam, 2022). This initiative aims to prepare teachers to deliver instruction in English, ostensibly to align with global educational standards. However, this policy appears unbalanced as it focuses solely on teachers' proficiency in English while neglecting the linguistic needs of students. The abrupt push for EMI without a comprehensive strategy to support both teachers and students is unlikely to reflect a genuine commitment to EMI. Instead, it highlights a discrepancy between government rhetoric and the realities within educational institutions.

To move beyond tokenism, it is essential to develop policies that are both comprehensive and inclusive. For e-learning, this means investing in the necessary technological infrastructure and ensuring students have access to the digital tools they need. Additionally, the implementation of EMI should be gradual and inclusive, incorporating comprehensive language training programmes for both teachers and students, and recognising the importance of multilingualism in the Algerian context. Furthermore, policymakers must engage in continuous dialogue with educators, students, and other stakeholders to ensure that policies are grounded in the realities of the educational environment. This approach will help bridge the gap between policy rhetoric and practice, fostering a more genuine and effective implementation of sustainability practices in HE.

6.3 Conclusion

In this chapter, we have discussed the multifaceted dimensions and challenges of implementing sustainability in Algerian HE. The historical review of HE reforms in Algeria unveils a landscape heavily influenced by political factors and characterised by a lack of consistent SD. Cooperation between France and Algeria in the educational sector (Matasci et al., 2020), along with reforms in the 70s, 80s, and 2000s, were less effective to establish a structured plan for long-term sustainability. Instead, these reforms mirrored socio-political changes without adequately integrating sustainability principles.

A significant barrier identified is the lack of technological infrastructure, which is particularly problematic for digital learning initiatives. Despite the emphasis on digitalisation raising expectations, many HEIs still lack the necessary resources to implement effective digital practices. Additionally, the introduction of EMI has posed its own set of challenges. The abrupt shift to English has marginalised non-English speakers and highlighted the need for a more balanced and gradual approach. A multilingual strategy that incorporates English while respecting the linguistic diversity of Arabic and French speakers is essential for fostering an inclusive educational environment.

Furthermore, the issue of tokenism in sustainability practices is prevalent. Policies often appear superficial, consisting of temporary measures that are less effective to address underlying issues. For example, the promotion of e-learning platforms and the push for EMI without adequate preparatory measures exemplify this tokenistic approach. To achieve SDG4 quality education, a comprehensive approach is necessary. Engaging stakeholders at all levels, from bottom to top-level decision-makers, is crucial for developing effective ESD reforms.

References

Bouherar, S., & Ghafsi, A. (2021). *Algerian languages in education: Conflicts and reconciliation*. Palgrave Macmillan. https://link.springer.com/book/10.1007/978-3-030-89324-8

Dale, A., & Newman, L. (2005). Sustainable development, education and literacy. *International Journal of Sustainability in Higher Education, 6*(4), 351–362.

Grimes, N. D., & Porter, W. (2023). Closing the digital divide through digital equity: The role of libraries and librarians. *Public Library Quarterly*. https://doi.org/10.1080/01616846.2023.2251348

Haouas, A., Ochi, A., & Labidi, M. A. (2024). Sources of Algeria's economic growth, 1979–2019: Augmented growth accounting framework and growth regression method. *Regional Science Policy & Practice, 16*(3), 1–19.

Jacob, C. (2020). English as a decolonial language: Academic frames, popular discourses & language practices in Algeria. *The Journal of North African Studies, 25*(6), 1013–1032.

Khandriche, M. (1996). Algeria and France: The enduring strength of cultural ties. In C. Alden & J. P. Daloz (Eds.), *Paris, Pretoria and the African Continent: The international relations of states and societies in transition* (pp. 54–62). Palgrave Macmillan.

Maraf, B., & Vanci Osam, U. (2022). The smile revolution (hirak) as a driving force for an English 'tidal wave' and foreign language policy-making in Algeria. *Current Issues in Language Planning, 24*(2), 179–200.

Matasci, D., Jerónimo, M. B., & Dores, H. G. (2020). Introduction: Historical trajectories of education and development in (post)colonial Africa. In D. Matasci, M. B. Jerónimo, & H. G. Dores (Eds.), *Education and development in colonial and postcolonial Africa: Policies, paradigms, and entanglements, 1890s–1980s* (pp. 1–28). Palgrave Macmillan.

Moore, J. (2005). Is higher education ready for transformative learning? A question explored in the study of sustainability. *Journal of Transformative Education, 3*(1), 76–91.

Piao, X., & Managi, S. (2023). The international role of education in sustainable lifestyles and economic development. *Scientific Reports, 13*, 1–12.

Scoffham, S., & Rawlinson, S. (2022). *Sustainability education: A classroom guide*. Bloomsbury.

Appendix A: Questionnaire

Digitalisation in Algerian higher education—الرقمنة في التعليم العال الجزائري

Introduction: Digitalisation in Algerian higher education is a transformative process that involves integrating digital technologies into various aspects of academic and administrative functions. This research explores the beliefs and attitudes of students and teachers surrounding the adoption of digitalisation in Algerian higher education institutions, shedding light on the challenges, opportunities, and perceptions of key stakeholders.

Procedures: If you agree to participate in this study, you will be asked to complete a questionnaire.

Risks and benefits: There are no anticipated physical risks associated with participation in this study. The potential benefits include contributing valuable insights to the ongoing discourse on digitalisation in Algerian higher education.

Confidentiality: Your identity will be kept confidential, and any information collected will be anonymised and used solely for research purposes. Only the principal investigator and research team will have access to the data.

Voluntary participation: Participation in this study is voluntary, and you have the right to withdraw at any time without providing a reason. Your decision to participate or withdraw will not affect your current or future relationship with the institution you are part of.

Contact information: If you have any questions or concerns about the study, you may contact the principal investigator, Salim Bouherar, at salim.bouherar@gmail.com, Setif 2 University.

Consent: I have read the information provided above, and I understand the purpose, procedures, risks, and benefits associated with participating in this research study. I agree to participate voluntarily.

مقدمة: الرقمنة في التعليم العالي الجزائري هي عملية تحويلية تنطوي على دمج التقنيات الرقمية في مختلف جوانب الوظائف الأكاديمية والإدارية. يستكشف هذا البحث معتقدات ومواقف الطلاب والمدرسين

S. Bouherar, S. Salem, *Sustainability in Algerian Higher Education*,
https://doi.org/10.1007/978-3-031-81347-4

المحيطة باعتماد الرقمنة في مؤسسات التعليم العالي الجزائرية، مع تسليط الضوء على التحديات والفرص وتصورات أصحاب المصلحة الرئيسيين.

الإجراءات: إذا وافقت على المشاركة في هذه الدراسة، فسوف يطلب منك ما يلي

إكمال الاستبيان

معلومات الاتصال: إذا كان لديك أي أسئلة أو مخاوف بشأن الدراسة، يمكنك الاتصال بالباحث الرئيسي، سليم بوهرار، علىالمشاركة التطوعية: المشاركة في هذه الدراسة تطوعية، ويحق لك الانسحاب في أي وقت دون إبداء الأسباب. لن يؤثر قرارك بالمشاركة أو الانسحاب على علاقتك الحالية أو المستقبلية مع المؤسسة التي أنت جزء منهاالسرية: سيتم الحفاظ على سرية هويتك، وسيتم إخفاء هوية أي معلومات تم جمعها واستخدامها فقط لأغراض البحث. لن يتمكن سوى الباحث الرئيسي وفريق البحث من الوصول إلى البياناتالمخاطر والفوائد: لا توجد مخاطر جسدية متوقعة مرتبطة بالمشاركة في هذه الدراسة. وتشمل الفوائد المحتملة المساهمة برؤى قيمة في الخطاب المستمر حول الرقمنة في التعليم العالي الجزائري .salim bouherar@gmail.com الموافقة: لقد قرأت المعلومات المقدمة أعلاه، وأفهم الغرض والإجراءات والمخاطر والفوائد المرتبطة بالمشاركة في هذه الدراسة البحثية. أوافق على المشاركة طوعاجامعة سطيف 2

1. Give your consent to proceed please—أعط موافقتك للمضي قدما من فضلك

I agree to participate voluntarily—أوافق على المشاركة طوعا

2. Are you—هل أنت

A teacher—أستاذ

A student—طالب

3. How old are you?—كم عمرك

18–25

26–30

31 35

36–40

41>

4. Where is your university located?—أين تقع جامعتك

Adrar

Chlef

Laghouat

Oum El Bouaghi

Batna

Bejaia

Biskra

Bechar

Blida

Bouira

Tamanrasset

Tebessa

Tlemcen

Tiaret

Tizi Ouzou

Algiers

Djelfa

Jijel

Setif

Saïda

Skikda
Sidi Bel Abbès
Annaba
Guelma
Constantine
Médéa
Mostaganem
M'Sila
Mascara
Ouargla
Oran
El Bayadh
Illizi
Bordj Bou Arréridj
Boumerdes
El Tarf
Tindouf
Tissemsilt
El Oued
Khenchela
Souk Ahras
Tipaza
Mila
Aïn Defla
Naâma
Aïn Témouchent
Ghardaïa
Relizane
Timimoun
Bordj Badji Mokhtar
Ouled Djellal
Béni Abbès
Ain Salah
Ain Guezzam
Touggourt
Djanet
El M'Ghair
El Menia

5. Do you have a personal computer (PC)?—هل لديك جهاز كمبيوتر شخصي؟
Yes—نعم
No—لا

6. Do you have access to Internet at home (Idoom ADSL or Fibre)? Do not include your 3G/4G phone network. If you answer yes, go to question 6, if you answer no, go to question 7—
هل لديك إمكانية الإتصال بالإنترنت في المنزل

(Idoom ADSL أو Fibre)
إذا كانت إجابتك بنعم فانتقل إلى السؤال رقم 6، وإذا كانت إجابتك لا فانتقل إلى السؤال رقم 7
Yes—نعم
No—لا

7. Do you think your Internet can be counted on when needed and does it have a good quality feature?—هل تعتقد أنه يمكن الاعتماد على الإنترنت الخاص بك عند الحاجة إليه وهل يتمتع بميزة الجودة الجيدة؟
Very good—جيد جدًا
Good—جيد
Fair—متوسط
Poor—ضعيف
Very poor—ضعيف جدا

8. The university provides good technological facilities in learning (e.g. good e-learning spaces, good Internet, data shows, quality libraries equipped with Internet and computers)—توفر الجامعة مرافق تكنولوجية جيدة في التعلم (مثل مساحات التعلم عن بعد الجيدة، والإنترنت الجيد، وعروض البيانات، والمكتبات عالية الجودة المجهزة بالإنترنت وأجهزة الكمبيوتر)
Strongly agree—بشدة
Agree—موافق
Not sure—متأكد
Disagree—موافق
Strongly disagree—غير موافق بشدة

9. How do you rate maintenance of any technical breakdown in your university?—كيف تقيم صيانة أي عطل فني في جامعتك
Very good—جيد جدًا
Good—جيد
Fair—متوسط
Poor—ضعيف
Very poor—ضعيف جدا

10. The Higher Education Ministry launched the aim to digitalise universities across the country, do you think this aim would serve more?—أطلقت وزارة التعليم العالي هدف رقمنة الجامعات في جميع أنحاء البلاد، هل تعتقد أن هذا الهدف سيخدم أكثر
Teaching/learning purposes—أغراض التدريس/التعلم
Administrative purposes—الأغراض الإدارية

11. What technological devices do you use for e-learning?—ما هي الأجهزة التكنولوجية التي تستخدمها للتعلم عن بعد؟
Laptop computer—محمول
Desk computer—مكتبي
iPad—اى باد
Smart phone—هاتف ذكي
No device at all—لا شئ

12. What are the elements you think the university should have to reinforce/enhance the e-learning experience?—ما هي العناصر التي تعتقد أنه ينبغي للجامعة أن تمتلكها لتعزيز تجربة التعلم عن بعد؟

13. Do you think your e-learning experience will qualify you globally to integrate in different e-learning experiences abroad in other developed countries?—هل تعتقد أن تجربتك في التعلم عن بعد سوف تؤهلك عالميًا للاندماج في تجارب التعلم عن بعد المختلفة بالخارج في الدول المتقدمة الأخرى؟

Strongly agree—بشدة
Agree—موافق
Not sure—متأكد
Disagree—موافق
Strongly disagree—غير موافق بشدة

14. How does your university mostly contact you about your academic issues (e.g. student lists, exam timetable, supervisory meeting, make up sessions, meetings, quizzes, disciplinary committee list, teachers' absences)?—كيف تتصل بك جامعتك في الغالب بشأن مشكلاتك الأكاديمية

مثال: قوائم الطلاب، جدول الامتحانات، اجتماعات المشرفين، المحاضرات التعويضية، الاجتماعات، (الاختبارات، قائمة اللجان التأديبية، غيابات الأساتذة...إلخ)

Emails—رسائل البريد الإلكتروني
E-learning platforms—منصات التعلم عن بعد
Facebook—فيسبوك

15. If it was up to you, what aspects would you change/include to enhance e-learning in your university?—إذا كان الأمر متروكًا لك، ما هي الجوانب التي ستغيرها/تدرجها لتعزيز التعلم عن بعد في جامعتك؟

16. How satisfied are you with the current e-learning experience in your university?—ما مدى رضاك عن تجربة التعلم عن بعد الحالية في جامعتك؟

Appendix B: Semi-structured Interview

Introduction: English as a Medium of Instruction (EMI) in Algerian higher education is a transformative process that involves integrating English to replace French in higher education. This research explores the beliefs and attitudes of students and teachers surrounding the adoption of EMI in Algerian higher education institutions, shedding light on the challenges, opportunities, and perceptions of key stakeholders.

Procedures: If you agree to participate in this study, you will be asked to complete this interview.

Risks and benefits: There are no anticipated physical risks associated with participation in this study. The potential benefits include contributing valuable insights to the ongoing discourse on EMI in Algerian higher education.

Confidentiality: Your identity will be kept confidential, and any information collected will be anonymised and used solely for research purposes. Only the principal investigator and research team will have access to the data.

Voluntary participation: Participation in this study is voluntary, and you have the right to withdraw at any time without providing a reason. Your decision to participate or withdraw will not affect your current or future relationship with members of the research team.

Contact information: If you have any questions or concerns about the study, you may contact the principal investigator, Salim Bouherar, at salim.bouherar@gmail.com OR s.bouherar@univ-setif2.dz, Setif 2 University.

Consent: I have read the information provided above, and I understand the purpose, procedures, risks, and benefits associated with participating in this research study. I agree to participate voluntarily.

S. Bouherar, S. Salem, *Sustainability in Algerian Higher Education*,
https://doi.org/10.1007/978-3-031-81347-4

Implementation of English as a Medium of Instruction (EMI) in Algeria

1. What does the choice of English as a medium of instruction signify to you?
2. What factors have contributed to the adoption of EMI in Algerian higher education institutions?
3. What are the primary aims behind the integration of EMI in Algerian higher education?
4. How would you describe the current status of EMI in Algeria?
5. How can educators balance the use of English with ensuring students have a solid understanding of content?
6. What challenges are encountered in implementing EMI in Algerian higher education, and how are these challenges being addressed?
7. Is there a risk of marginalisation associated with prioritising English, and how could it be mitigated?
8. Are there specific opportunities (e.g. employment, research) that arise from using EMI in Algerian higher education?
9. How are teachers prepared to effectively deliver their subjects in English?
10. Are there any initiatives or programmes designed to support teachers and students facing linguistic challenges in adapting to EMI?
11. What collaborations (e.g. scholarships and training for educators) exist to ensure the successful implementation of EMI?
12. In what way has the implementation of EMI influenced the academic performance of teachers and students in Algeria?

The Future of English as a Medium of Instruction (EMI) in Algeria

1. How do you see the future of EMI in Algeria, and what measures are being taken to continuously improve and refine the implementation process?
2. How is the implementation of EMI ensuring equitable and inclusive access to vocational and educational opportunities?
3. What strategies can teachers employ to establish safe learning environments for students with diverse language backgrounds?
4. To what extent does EMI in Algeria foster students' ability to coexist globally?

References

Abbas, A. (2023). The requirements of the directive plan for the digitization of the higher education sector: Reality and hope. *Journal of Legal and Economic Research, 6*(3), 271–283.

Abdulrazak, F. (1982). Arabization in Algeria. *MELA Notes, 26*, 22–43.

Abid, M., Benmeriem, M., Gheraia, Z., Sekrafi, H., Abdelli, H., & Meddah, A. (2023). Asymmetric effects of economy on unemployment in Algeria: Evidence from a nonlinear ARDL approach. *Cogent Economics & Finance, 11*(1), 1–20.

Abubakar, I. R., Aina, Y. A., & Alshuwaikhat, H. M. (2020). Sustainable development at Saudi Arabian universities: An overview of institutional frameworks. *Sustainability, 12*(19), 1–15.

Aburizaizah, S. J. (2022). The role of quality assurance in Saudi higher education institutions. *International Journal of Educational Research Open, 3*, 1–8. https://doi.org/10.1016/j.ijedro.2022.100127

Adbullah, M. (2020). UAE education heads in the path of sustainability. *The Sustainabilist.* https://thesustainabilist.ae/uae-education-heads-in-the-path-of-sustainability/

Agasisti, T., & Bertoletti, A. (2022). Higher education and economic growth: A longitudinal study of European regions 2000–2017. *Socio Economic Planning Science, 81*. https://doi.org/10.1016/j.seps.2020.100940

Airey, J. (2016). EAP, EMI or CLIL? In K. Hyland & P. Shaw (Eds.), *The Routledge handbook of English for academic purposes* (pp. 71–83). Routledge.

Aïssaoui, A. M. (2001). *Algeria: The political economy of oil and gas.* Oxford University Press.

Alahmari, M., Issa, T., Issa, T., & Nau, S. Z. (2019). Faculty awareness of the economic and environmental benefits of augmented reality for sustainability in Saudi Arabian universities. *Journal of Cleaner Production, 226*, 259–269.

Aleixo, A. M., Leal, S., & Azeiteiro, U. M. (2016). Conceptualization of sustainable higher education institutions, roles, barriers, and challenges for sustainability: An exploratory study in Portugal. *Journal of Cleaner Production, 172*, 1664–1673.

S. Bouherar, S. Salem, *Sustainability in Algerian Higher Education*, https://doi.org/10.1007/978-3-031-81347-4

Alexander, C. (2002). The architecture of militancy: Workers and the state in Algeria, 1970–1990. *Comparative Politics, 34*(3), 315–335.

Alfaki, I., & Ahmed, A. (2016). *From oil to knowledge: Transforming the United Arab Emirates into a knowledge-based economy.* Routledge.

Algeria Press Service. (2022). Presidents Tebboune, Macron chair signing ceremony of 5 bilateral cooperation agreements. *Online article.* https://www.aps.dz/en/algeria/44720-presidents-tebboune-macron-chair-signing-ceremony-of-5-bilateral-cooperation-agreements

Algeria: France-Algeria independence agreements (Evian Agreements). (1962). *International Legal Materials, 1*(2), 214–230. http://www.jstor.org/stable/20689578

Algerie Telecom. (2024). *Message from the president and CEO.* https://www.algerie-telecom.dz/en/page/mot-du-president-directeur-general-p141

Alhassan, A. (2021). Challenges and professional development needs of EMI lecturers in Omani higher education. *Sage Open, 11*(4), 21582440211061527.

Alismaiel, O. A. (2021). Using structural equation modeling to assess online learning systems' educational sustainability for university students. *Sustainability, 13*(24), 13565.

Alkhateeb, H., & Bouherar, S. (2022). Sustainability and educational language policy in Arab higher education: Findings from Q research. *Current Issues in Language Planning, 23*(5), 1–23.

Alkhayyal, B., Labib, W., Alsulaiman, T., & Abdelhadi, A. (2019). Analyzing sustainability awareness among higher education faculty members: A case study in Saudi Arabia. *Sustainability, 11*(23), 1–13.

Al-Musalmani, F., & Maalouf, S. (2023). Qatar Foundation: A sustainability innovator. In L. Cochrane & R. Al-Hababi (Eds.), *Sustainable Qatar: Social, political and environmental perspectives* (pp. 73–92). Springer.

Al-Thani, W. A., Ari, I., & Koç, M. (2021). Education as a critical factor of sustainability: Case study in Qatar from the teachers' development perspective. *Sustainability, 13*(20), 1–32.

An, Y. J., & Reigeluth, C. (2011). Creating technology-enhanced, learner-centered classrooms. *Journal of Digital Learning in Teacher Education, 28*(2), 54–62.

Arabi, M. (2018). Adoption of (LMD) system in Algerian higher education institutions: Response of international trends or a quality culture of higher education? *Innovation Creativity, 8*(1), 230–244.

Assalihee, M., & Boonsuk, Y. (2022). Reforming religion-based higher education for sustainable development: The case of Islamic studies international program at Prince of Songkla University. In M. Öztürk (Ed.), *Engagement with sustainable development in higher education: Universities as transformative spaces for sustainable futures* (pp. 67–82). Springer.

Atheer. (2023). Qira'at fi almashhad aljazayri mae wazir alkharijia aljazayri Ahmad Attaf – budkast dhu shaan (A reading of the Algerian scene with the Algerian Foreign Minister Ahmed Attaf – dhu shaan podcast). *YouTube Video.* https://www.youtube.com/watch?v=Yd-i-EJYZBw&t=1164s

Ayres, R. U., Turton, H., & Casten, T. (2007). Energy efficiency, sustainability and economic growth. *Energy, 32*(5), 634–648.

Bader, E. E. (2005). Sustainable hotel business practices. *Journal of Retail & Leisure Property, 5*(1), 70–77.

Baker, W., & Fang, F. (2021). 'So maybe I'm a global citizen': Developing intercultural citizenship in English medium education. *Language, Culture and Curriculum, 34*(1), 1–17.

Bälter, O., Kann, V., Mutimukwe, C., & Malmström, H. (2023). English-medium instruction and impact on academic performance: A randomized control study. *Applied Linguistics*. https://doi.org/10.1515/applirev-2022-0093

Bandarage, A. (2013). *Sustainability and well-being: The middle path to environment, society and the economy*. Palgrave Macmillan.

Banegas, D. L., & Busleimán, G. M. (2021). EMI materials in online initial English language teacher education. In D. Lasagabaster & A. Doiz (Eds.), *Language use in English-medium instruction at university: International perspectives on teacher practice* (pp. 100–125). Routledge.

Bansal, P., & DesJardine, M. R. (2014). Business sustainability: It is about time. *Strategic Organization, 12*(1), 70–78.

Bélaïd, F., & Abderrahmani, F. (2013). Electricity consumption and economic growth in Algeria: A multivariate causality analysis in the presence of structural change. *Energy Policy, 55*, 286–295.

Bellatrache, R., & Aloutti, L. (2020). E-learning challenges in higher education institutions: Algeria case study. *Journal of Economic and Human Development, 11*(2), 264–276.

Belmihoub, K. (2018). English in a multilingual Algeria. *World Englishes, 37*(2), 207–227.

Benchicou, S., Aichouni, M., & Nehari, D. (2010). E-learning in engineering education: A theoretical and empirical study of the Algerian higher education institution. *European Journal of Engineering Education, 35*(3), 325–343.

Benchikh, M. (2017). Governance of higher education institutions in Algeria: Inventory and assessment. In A. Farazmand (Ed.), *Global encyclopedia of public administration, public policy, and governance* (pp. 1–7). Springer.

Bennoune, M. (2000). *Education, culture et droeloppement en Algerie: Bilan et perspectives du systeme educatif*. Marinoor-ENAG.

Benouar, D. (2013). Algerian experience in education, research and practice. *Procedia – Social and Behavioral Sciences, 102*, 361–367.

Benrabah, M. (2005). Language planning situation in Algeria. *Current Issues in Language Planning, 6*(4), 379–502.

Benrabah, M. (2013). *Language conflict in Algeria: From colonialism to post-independence*. Multilingual Matters.

Bentahar, Z. (2020). "*Ytnahaw ga'*!": Algeria's cultural revolution and the role of language in the early stages of the spring 2019 Hirak. *Journal of African Cultural Studies, 33*(4), 471–488.

Benziane, I., & Boutaba, N. (2015). Waqie tatbiq nizam l.m.d fi aljamieat aljazayiria: dirasa maydania liwijhat nazar al'iidariiyn fi eayinat min aljamieat (The reality of implementing the LMD system in Algerian universities: A field study of administrators point of views in a sample of universities). *The Social and Human Sciences Review, 8*(2), 63–83.

Bessaid, O. N., & Benyahia, C. (2023). The role of e-learning in the Algerian Open University to achieve the development of human capital. In S. Sedkaoui, M. Khelfaoui, R. Benaichouba, & K. M. Belkebir (Eds.), *International conference on managing business through wen analytics* (pp. 167–178). Springer.

Bint El Hassan, S., Cordova, K. E., Rabadi, G., & Abu Elhaija, W. (2021). The (un) sustainability of higher education institutions in Jordan. *Frontiers in Sustainability, 2*, 1–6.

Birnbacher, D., & Thorseth, M. (2015). *The politics of sustainability: Philosophical perspectives.* Routledge.

Boeve-de Pauw, J., Gericke, N., Olsson, D., & Berglund, T. (2015). The effectiveness of education for sustainable development. *Sustainability, 7*(11), 15693–15717.

Bouchrika, I., Harrati, N., Wanick, V., & Wills, G. (2021). Exploring the impact of gamification on student engagement and involvement with e-learning systems. *Interactive Learning Environments, 29*(8), 1244–1257.

Boudiaf, S. M., & Zeghichi, S. (2021). New, smart, and sustainable cities in Algeria: Realities and challenges. Case of the city of Sidi Abdellah in Algiers region. In F. Trapani, N. Mohareb, F. Rosso, D. Kolokotsa, S. Maruthaveeran, & M. Ghoneem (Eds.), *Advanced studies in efficient environmental design and city planning* (pp. 209–224). Springer.

Bouherar, S., & Ghafsi, A. (2021). *Algerian languages in education: Conflicts and reconciliation.* Palgrave Macmillan. https://link.springer.com/book/10.1007/978-3-030-89324-8

Branka, D. M., & Molino, A. (2022). Translanguaging in EMI: Lecturers' perspectives and practices. *Ricognizioni, 9*(17), 29–47.

Braun, V., & Clarke, V. (2006). Using thematic analysis in psychology. *Qualitative Research in Psychology, 3*(2), 77–101.

Brinkmann, R. (2020). *Environmental sustainability in a time of change* (Vol. 279). Palgrave Macmillan.

Brinton, D., Snow, M. A., & Wesche, M. B. (2003). *Content-based second language instruction.* University of Michigan Press.

British Council. (2014). *Algerians to study in UK.* https://opportunities-insight.britishcouncil.org/news/market-news/algerians-study-uk

British Council. (2021). Signing of the Memorandum – Higher Education and Scientific Research. *Online article.* https://www.britishcouncil.dz/en/programmes/education/higher-education-commission

Bruder, J., Hamwy, N., Sellami, A., & Romanowski, M. H. (2023). Examining Qatar's private and public-school teachers' readiness for teaching about sustainability. *Globalisation, Societies and Education, 21*(4). https://doi.org/10.1080/14767724.2023.2262404

Bruton, A. (2011). Is CLIL so beneficial, or just selective? Re-evaluating some of the research. *System, 39*(4), 523–532.

Bruton, A. (2013). CLIL: Some of the reasons why… and why not. *System, 41*(3), 587–597.

Bryman, A. (2016). *Social research methods* (5th ed.). Oxford University Press.

Bude, U. (1983). The adaptation concept in British colonial education. *Comparative Education, 19*(3), 341–355.

Byram, M. (1997). *Teaching and assessing intercultural communicative competence.* Multilingual Matters LTD.

Canagarajah, S. (2011). Codemeshing in academic writing: Identifying teachable strategies of translanguaging. *The Modern Language Journal, 95*(3), 401–417.

Cardoni, A., & Kiseleva, E. (2023). *Sustainable governance: Concepts, metrics and contexts.* Springer.

Carrió Pastor, M. L., & Perry, D. (2010). The collaborative approach in content and language integrated learning. *Revista Alicantina de Estudios Ingleses, 23*, 69–81.

Carrió-Pastor, M. L. (2021). CLIL vs EMI: different approaches or the same dog with a different collar? In M. L. Carrió-Pastor & B. Bellés Fortuño (Eds.), *Teaching language and content in multicultural and multilingual classrooms: CLIL and EMI approaches* (pp. 13–30). Palgrave Macmillan.

Chankseliani, M., & McCowan, T. (2021). Higher education and the sustainable development goals. *Higher Education, 81*(1), 1–8.

Christensen, H. B., Hail, L., & Leuz, C. (2021). Mandatory CSR and sustainability reporting: Economic analysis and literature review. *Review of Accounting Studies, 26*(3), 1176–1248.

Cochrane, L., & Al-Hababi, R. (Eds.). (2023). *Sustainable Qatar: Social, political and environmental perspectives*. Springer.

Colonna, F. (2004). 'What has become of my friends?' Questions regarding the new Algerian 'diaspora' in France. *The Journal of North African Studies, 9*(2), 126–139.

Costa, F., & Mastellotto, L. (2022). The role of English for specific purposes (ESP) in supporting the linguistic dimension in English-medium instruction (EMI). *CLIL: Journal of Innovation and Research in Plurilingual and Pluricultural Education, 5*(2), 37–52.

Coyle, D., Hood, P., & Marsh, D. (2010). *Content and language integrated learning*. Cambridge University Press.

D'amato, S., & Panta, G. D. (2017). Does rentierism have a conditional effect on violence? Regime oil dependency and civil war in Algeria. *The Extractive Industries and Society, 4*(2), 361–370.

Dafouz, E. (2018). English-medium instruction in multilingual university settings – An opportunity for developing language awareness. In P. Garrett & J. M. Cots (Eds.), *The Routledge handbook of language awareness* (1st ed., pp. 170–185). Routledge.

Dafouz, E. (2021). Crossing disciplinary boundaries: English-medium education (EME) meets English for Specific Purposes (ESP). *Ibérica, 41*, 13–38.

Daghbouche, N. (2011). The impact of the credit transfer accumulation system on Algeria. *The Journal of North African Studies, 16*(3), 465–470.

Dale, A., & Newman, L. (2005). Sustainable development, education and literacy. *International Journal of Sustainability in Higher Education, 6*(4), 351–362.

Daoudi, A. (2018). Multilingualism in Algeria: Between 'soft power', 'Arabisation', 'Islamisation', and 'globalisation'. *The Journal of North African Studies, 23*(3), 460–481.

Davidson, M. (2010). Social sustainability and the city. *Geography Compass, 4*(7), 872–880.

Davis, M. H. (2019). The layers of history beneath Algeria's protests. *Current History, 118*(812), 337–342.

De Fine Licht, K., & Folland, A. (2019). Defining "social sustainability": Towards a sustainable solution to the conceptual confusion. *Etikk i Praksis-Nordic Journal of Applied Ethics, 2*, 21–39.

Dearden, J. (2014). *English as a medium of instruction – A growing global phenomenon*. British Council. https://www.britishcouncil.es/sites/default/files/british_council_english_as_a_medium_of_instruction.pdf

Dempsey, N., Bramley, G., Power, S., & Brown, C. (2011). The social dimension of sustainable development: Defining urban social sustainability. *Sustainable Development, 19*(5), 289–300.

Denscombe, M. (2007). *The good research guide for small-scale social research projects* (3rd ed.). Open University Press.

Didham, R. (2018). *Education for sustainable development and the SDGs: Learning to act, learning to achieve. Policy brief; Advancing ESD policy, UNESCO global action program on education for sustainable development.* https://en.unesco.org/sites/default/files/gap_pn1_-_esd_and_the_sdgs_policy_brief_4_page_version.pdf

Diener, E., & Crandall, R. (1978). *Ethics in social and behavioral research.* the University of Chicago Press.

Disterheft, A., da Silva Caeiro, S. S. F., Ramos, M. R., & de Miranda Azeiteiro, U. M. (2012). Environmental Management Systems (EMS) implementation processes and practices in European higher education institutions – Top-down versus participatory approaches. *Journal of Cleaner Production, 31*, 80–90.

Dreer, B. (2022). Teacher well-being: Investigating the contributions of school climate and job crafting. *Cogent Education, 9*(1), 1–13.

Dreher, A., Lamla, M. J., Lein, S. M., & Somogyi, F. (2009). The impact of political leaders' profession and education on reforms. *Journal of Comparative Economics, 37*(1), 169–193.

Dris-Ait Hamadouche, L., Oussedik, F., & Taleb Ibrahimi, K. (Eds.). (2022). *L'université désacralisée: Recul de l'éthique et explosion de la violence.* Koukou.

Durham, B. (2020). *Une aventure sociale et humaine*: The Service *des Centres Sociaux* in Algeria, 1955–1962. In D. Matasci, M. B. Jerónimo, & H. G. Dores (Eds.), *Education and development in colonial and postcolonial Africa: Policies, paradigms, and entanglements, 1890s–1980s* (pp. 55–82). Palgrave Macmillan.

Ekins, P. (2011). Environmental sustainability: From environmental valuation to the sustainability gap. *Progress in Physical Geography, 35*(5), 629–651.

Elbiled.net. (2018). Hadha huwa adad altalaba aljazayiriiyn fi aljamieat alfaransia (This is the number of Algerian students in French universities). *Online article.* https://www.elbilad.net/derniere- info/70542-هذا-هو-عدد-الطلبة-الجزائريين-في-الجامعات-الفرنسية

Elliott, S. R. (2005). Sustainability: An economic perspective. *Resources, Conservation and Recycling, 44*(3), 263–277.

ENTV. (2023). liqa' khas mae wazir altaelim alali aleilmii alsayid kamal bidari (A special meeting with the Minister of Higher Scientific Education, Mr. Kamal Badari). *YouTube Video.* https://www.youtube.com/watch?v=-vz432me-_o

Etikan, I., Musa, S. A., & Alkassim, R. S. (2016). Comparison of Convenience Sampling and Purposive Sampling. *American Journal of Theoretical and Applied Statistics, 5*(1), 1–4.

Ezeofor, I., & Lent, R. W. (2014). Social cognitive and self-construal predictors of well-being among African college students in the US. *Journal of Vocational Behavior, 85*(3), 413–421.

Faradji, M. A. (2015). 3rd World Conference on Psychology and Sociology, WCPS-2014: Sociological research in Algeria: Between theoretical language and social reality. *Procedia – Social and Behavioral Sciences, 185*, 352–356.

Fatahi, S., & Moradi, H. (2016). A fuzzy cognitive map model to calculate a user's desirability based on personality in e-learning environments. *Computers in Human Behavior, 63*, 272–281.

Ferguson, T., & Roofe, C. G. (2020). SDG 4 in higher education: Challenges and opportunities. *International Journal of Sustainability in Higher Education, 21*(5), 959–975.

Fia, M., Ghasemzadeh, K., & Paletta, A. (2023). How higher education institutions walk their talk on the 2030 agenda: A systematic literature review. *Higher Education Policy, 36*(3), 599–632.

Fischer, D., Jenssen, S., & Tappeser, V. (2015). Getting an empirical hold of the sustainable university: A comparative analysis of evaluation frameworks across 12 contemporary sustainability assessment tools. *Assessment & Evaluation in Higher Education, 40*(6), 785–800.

Foucault, M. (1991). Governmentality. In G. Burchell, C. Gordon, & P. Miller (Eds.), *The Foucault effect: Studies in governmentality* (pp. 87–104). The University of Chicago Press.

Galán-Muros, V. (2023). *General guidelines for the implementation of sustainability in higher education institutions.* https://www.un.org/sites/un2.un.org/files/unai_and_unesco_iesalc_guidelines_for_sdgs_implementation_final_version.pdf

Garcia, N. (2014). Tensions between cultural and utilitarian dimensions of language: A comparative analysis of 'multilingual' education policies in France and Germany. *Current Issues in Language Planning, 16*(1-2), 43–59.

García-Feijoo, M., Eizaguirre, A., & Rica-Aspiunza, A. (2020). Systematic review of sustainable-development-goal deployment in business schools. *Sustainability, 12*(1), 1–19.

Gardinier, D. E. (1988). Les recommandations de la Conférence de Brazzaville sur les problèmes d'éducation. In Institut Charles-de-Gaulle (Ed.), *Brazzaville, janvier-février 1944: aux sources de la décolonisation* (pp. 170–180). Plon.

Ghanem, D. (2022). *Understanding the persistence of competitive authoritarianism in Algeria.* Palgrave Macmillan.

Gherbi, M. (2012). Problematic of environment protection in Algerian cities. *Energy Procedia, 18*, 265–275.

Ghouati, A. (2022). Réformer l'enseignement supérieur par la professionnalisation des formations? Comparaison de trois expériences en Algérie. In M. Benguerna & J. B. Meyer (Eds.), *L'Université et ses acteurs dans un monde en turbulence* (pp. 1–20). CREAD-CRASC.

Gichuhi, L., Obiero, C., & Kangethe, M. (2022). Developing a monitoring and evaluation framework for education systems in view of sustainable development goals: A case from Kenya. In M. Öztürk (Ed.), *Educational response, inclusion and empowerment for SDGs in emerging economies* (pp. 39–50). Springer.

Giddings, B., Hopwood, B., & O'brien, G. (2002). Environment, economy and society: Fitting them together into sustainable development. *Sustainable Development, 10*(4), 187–196.

Goedegebuure, L., & Hayden, M. (2007). Overview: Governance in higher education – Concepts and issues. *Higher Education Research & Development, 26*(1), 1–11.

González-Torre, P. L., & Suárez-Serrano, E. (2022). A framework for implementing and reporting United Nations sustainable development goals in Spanish higher education institutions. *International Journal of Sustainability in Higher Education, 23*(6), 1349–1365.

Goodland, R. (1995). The concept of environmental sustainability. *Annual Review of Ecology and Systematics, 26*(1), 1–24.

Gopinathan, S. (2007). Globalisation, the Singapore developmental state and education policy: A thesis revisited. *Globalisation, Societies and Education, 5*(1), 53–70.

Government Budget. (2023). *Sectoral performance.* https://www.my.gov.sa/wps/portal/snp/aboutksa/governmentBudget/!ut/p/z0/04_Sj9CPykssy0xPLMnMz0vMAfIjo8zifT2dPQ38TQz9DUzDjA0CHV39TEy8XIwMzA30g1Pz9AuyHRUBPEIqgQ!!/

Government Communications Office. (2023). *Qatar National Vision 2030.* https://www.gco.gov.qa/en/about-qatar/national-vision2030/

Gradillas, M., & Thomas, L. D. W. (2023). Distinguishing digitization and digitalization: A systematic review and conceptual framework. *Journal of Product Innovation Management*. https://doi.org/10.1111/jpim.12690

Grandguillaume, G. (1997). Le Maghreb confronte a l' islamisme: Arabisation et demagogie en Algerie. *Le Monde Diplomatique*. https://www.monde-diplomatique.fr/1997/02/GRANDGUILLAUME/4593

Griebeler, J. S., Brandli, L. L., Salvia, A. L., Leal Filho, W., & Reginatto, G. (2022). Sustainable development goals: A framework for deploying indicators for higher education institutions. *International Journal of Sustainability in Higher Education, 23*(4), 887–914.

Griessler, E., & Littig, B. (2005). Social sustainability: A catchword between political pragmatism and social theory. *International Journal for Sustainable Development, 8*(1/2), 65–79.

Grimes, N. D., & Porter, W. (2023). Closing the digital divide through digital equity: The role of libraries and librarians. *Public Library Quarterly*. https://doi.org/10.1080/01616846.2023.2251348

Guba, E. G. (1981). Criteria for assessing the trustworthiness of naturalistic inquiries. *ECTJ, 29*(2), 75–91.

Guessoum, A. (2019). Introducing quality assurance in Algerian higher education: The case of the University of Science and Technology Houari Boumediene. In A. Badran, E. Baydoun, & J. R. Hillman (Eds.), *Major challenges facing higher education in the Arab World: Quality assurance and relevance* (pp. 335–352). Springer.

Gulf News. (2017). *UAE launches National Strategy for Higher Education 2030*. https://gulfnews.com/uae/education/uae-launches-national-strategy-for-higher-education-2030-1.2097191#

Gumperz, J. J. (1964). Linguistic and social interaction in two communities. *American Anthropologist, 66*(6), 137–153.

Hadakha, A. (2022). *khuta khumasiat bayn aljazayir wa alsin.. madha sayajni albaladan min altaeawun al'iistiratiji?* (A five-year plan between Algeria and China. What will the two countries gain from strategic cooperation?). https://www.aljazeera.net/ebusiness/2022/11/15/خطة-خماسية-بين-الجزائر-والصين-ماذا

Hadjam, L., & Tari, S. (2019). Sustainable development in Algeria: Analytical reading of the concept and constraints. *Revue Recherches et Etudes en Développement, 06*(02), 121–140.

Hanafi, S., & Arvanitis, R. (2016). *Knowledge production in the Arab World: The impossible promise*. Routledge.

Hanushek, E. A., & Wößmann, L. (2010). Education and economic growth. In P. Peterson, E. Baker, & B. McGaw (Eds.), *International encyclopedia of education* (3rd ed., pp. 245–252). Elsevier.

Haouas, A., Ochi, A., & Labidi, M. A. (2024). Sources of Algeria's economic growth, 1979–2019: Augmented growth accounting framework and growth regression method. *Regional Science Policy & Practice, 16*(3), 1–19.

Harbi, M. (2002). Le poids de l'histoire: Et la violence vint a l'Algerie. *Le Monde Diplomatique*. https://www.monde-diplomatique.fr/2002/07/HARBI/9218

Harmon, J. C. (2007). Let them use the internet: Why college instructors should encourage student internet use. *College Teaching, 55*(1), 2–4.

Heinberg, R., & Lerch, D. (2010). What is sustainability? *The Post Carbon Reader, 11*, 19. https://mycourses.aalto.fi/pluginfile.php/1159956/mod_page/content/9/Heinberg_WhatIsSustainability.pdf

Herreman, P. (1965). LES RELATIONS S'AGGRAVENT entre le régime Boumediene et le parti communiste français. *Le Monde*. https://www.lemonde.fr/archives/article/1965/09/13/les-relations-s-aggravent-entre-le-regime-boumediene-et-le-parti-communiste-francais_2201575_1819218.html

Hillman, J. R., & Baydoun, E. (2019). Quality assurance and relevance in academia: A review. In A. Badran, E. Baydoun, & J. R. Hillman (Eds.), *Major challenges facing higher education in the Arab World: Quality assurance and relevance* (pp. 13–68). Springer.

Holst, J. (2023). Towards coherence on sustainability in education: A systematic review of Whole Institution Approaches. *Sustainability Science, 18*, 1015–1030.

Hutchinson, T., & Waters, A. (1987). *English for specific purposes: A learning-centered approach*. Cambridge University Press.

Hüttner, J., & Smit, U. (2014). CLIL (Content and Language Integrated Learning): The bigger picture. A response to: A. Bruton. 2013. CLIL: Some of the reasons why... and why not. System 41 (2013): 587–597. *System, 44*, 160–167.

Idri, N. (2012). Education and reform to reach autonomous learners: Between reality and myth. *Procedia – Social and Behavioral Sciences, 69, 2174–2183.*

Ikerd, J. E. (2012). *The essentials of economic sustainability*. Kumarian Press.

Im, J. H., & Kim, J. (2015). Use of blended learning for effective implementation of English-medium instruction in a non-English higher education context. *International Education Studies, 8*(11), 1–15.

Jacob, C. (2019). 'Back to the "futur"': Mobility and immobility through English in Algeria. *Language & Communication, 68*, 6–16.

Jacob, C. (2020). English as a decolonial language: Academic frames, popular discourses & language practices in Algeria. *The Journal of North African Studies, 25*(6), 1013–1032.

Jeronen, E. (2022). Sustainable education. In S. Idowu, R. Schmidpeter, N. Capaldi, L. Zu, M. Del Baldo, & R. Abreu (Eds.), *Encyclopedia of Sustainable Management* (pp. 3488–3497). Springer.

Jorgensen, T. (2019). Universities move to achieve the SDGs-and approach the next hurdle. *Implementing the 2030 Agenda at Higher Education Institutions: Challenges and Responses*, pp. 44–89. https://www.hes-so.ch/data/documents/GLOBAL-UNIVERSITY-NETWORK-FOR-INNOVATION-12023.pdf

Jose, S., & Chacko, J. (2017). Building a sustainable higher education sector in the UAE. *International Journal of Educational Management, 31*(6), 752–765.

Kabel, A. (2023). The neoliberal linguistic consensus: Neoliberal multilingualism and linguistic governmentality in Morocco. *The Journal of North African Studies, 28*(4), 788–814.

Kadi, L. (2004). La politique linguistique Algerienne: Vers un etat des lieux. *Revue d'Amenagement Linguistique, 107*, 133–144.

Kadri, A. (2014a). Les conditions d'imposition du système scolaire français à l'Algérie. In A. Kadri (Ed.), *Instituteurs et enseignants en Algérie 1945–1975: Histoire et mémoires* (pp. 19–53). Karthala.

Kadri, A. (2014b). L'université d'Alger. In A. Kadri (Ed.), *Instituteurs et enseignants en Algérie 1945–1975: Histoire et mémoires* (pp. 175–185). Karthala.

Karaarslan-Semiz, G. (2022). Outdoor education for sustainability with systems thinking perspective. In G. Karaarslan-Semiz (Ed.), *Education for sustainable development in primary and secondary schools: Pedagogical and practical approaches for teachers* (pp. 39–53). Springer.

Karakhanyan, S. (2019). Quality assurance in the Arab region in the era of customization: Where do we stand in terms of relevance? In A. Badran, E. Baydoun, & J. R. Hillman (Eds.), *Major challenges facing higher education in the Arab World: Quality assurance and relevance* (pp. 211–224). Springer.

Karatsiori, M. (2023). In the pursuit of "Quality Education": From ancient times to the digital era, can there be a consensus? *Cogent Education, 10*(2), 1–33.

Karatzoglou, B. (2013). An in-depth literature review of the evolving roles and contributions of universities to education for sustainable development. *Journal of Cleaner Production, 49*, 44–53.

Kaur, J. (2020). Using English for interaction in the EMI classroom: Experiences and challenges in a Malaysian public university. In H. Bowles & A. C. Murphy (Eds.), *English-medium instruction and the internationalization of universities* (pp. 129–154). Palgrave Macmillan.

Kessar, S., Rabab'Ah, G., Al-Khadra, W., & Hamdan, H. J. (2021). The representation of the Algerian Hirak protest movement in the international media: France 24 and Al-Jazeera. *Cogent Social Sciences, 7*(1), 1–18.

Kestin, T., van den Belt, M., Denby, L., Ross, K., Thwaites, J., & Hawkes, M. (2017). *Getting started with the SDGs in universities: A guide for universities, higher education institutions, and the academic sector* (pp. 1–52). https://ap-unsdsn.org/wp-content/uploads/University-SDG-Guide_web.pdf

Khandriche, M. (1996). Algeria and France: The enduring strength of cultural ties. In C. Alden & J. P. Daloz (Eds.), *Paris, Pretoria and the African Continent: The international relations of states and societies in transition* (pp. 54–62). Palgrave Macmillan.

Khayati, A., & Selim, M. (2019). The status of innovation in Saudi universities. *Cogent Education, 6*(1), 1–12.

Khedouci, R. (2015). *Al madrasa wa al islah: 100 yawm fi al lajna al wataniya li islah al mandouma al terbawiya* [School and reform: 100 days in the national committee of educational reform]. Dar Alhadara.

Khelfaoui, H. (2003). Le champ universitaire Algérien entre pouvoirs politiques et champ économique. *Actes de la Recherche en Sciences Sociales, 3*(148), 34–46.

Khodr, H. (2011). The dynamics of international education in Qatar: Exploring the policy drivers behind the development of education city. *Journal of Emerging Trends in Educational Research and Policy Studies, 2*(6), 514–525.

Kilavuz, M. T., Grewal, S., & Kubinec, R. (2023). Ghosts of the Black Decade: How legacies of violence shaped Algeria's Hirak protests. *Journal of Peace Research, 60*(1), 9–25.

Kioupi, V., & Voulvoulis, N. (2020). Sustainable development goals (SDGs): Assessing the contribution of higher education programmes. *Sustainability, 12*(17). https://doi.org/10.3390/su12176701

Kodache, W. (2023). *Aljamieat aljazayiria ealaa bued khatwa min al alamia* [The Algerian university is a step away from internationalism]. https://elikhbaria.dz/الجامعة-الجزائرية-على-بعد-خطوة-من-العا/

Korean Development Institute. (2013). *Establishment of Algeria's national vision 2030*. Korean Development Institute.

Kormos, J. (2017). *The second language learning processes of students with specific learning difficulties*. Routledge.

Kouachi, M., Sergui, J., & Rami, K. (2019). The reality of higher education in Algeria within the framework of the reform program. *Journal of Studies in Economics and Management, 2*(1), 299–316.

Kramsch, C. (1998). *Language and culture.* Oxford University Press.

Kumar, R. (2011). *Research methodology: A step-by-step guide for beginners* (3rd ed.). Sage.

Lagha, H., & Bachi, A. (2018). Sustainable development in Algeria. *Algerian Journal of Environmental Science and Technology, 4*(2), 742–749.

Larimian, T., & Sadeghi, A. (2021). Measuring urban social sustainability: Scale development and validation. *Environment and Planning B: Urban Analytics and City Science, 48*(4), 621–637.

Laskaris, E. (2016). *La coopération culturelle franco-algérienne: Les coopérants français en Algérie indépendante: Enseignants des écoles (instituteurs et professeurs) et universitaires (1962–1980).* Doctoral thesis, Université Paris-Est. ProQuest Dissertations and Theses Global. https://theses.hal.science/tel-01691333

Lassoued, Z., Alhendawi, M., & Bashitialshaaer, R. (2020). An exploratory study of the obstacles for achieving quality in distance learning during the COVID-19 pandemic. *Education Science, 10*(9), 1–13.

Layachi, A. (2001). The private sector in the Algerian economy: Obstacles and potentials for a productive role. *Mediterranean Politics, 6*(2), 29–50.

Le Hoang Ngo, P. (2021). Translanguaging practices in EMI settings from the perspective of student agency: An example from Vietnamese higher education. In W. Tsou & W. Baker (Eds.), *English-medium instruction translanguaging practices in Asia: Theories, frameworks and implementation in higher education* (pp. 121–141). Springer.

Le Roux, C. S. (2017). Language in education in Algeria: A historical vignette of a 'most severe' sociolinguistic problem. *Language & History, 60*(2), 112–128.

Leal Filho, W., Frankenberger, F., Salvia, A. L., Azeiteiro, U., Alves, F., Castro, P., Will, M., Platje, J., Lovren, V. O., Brandli, L., Price, E., Doni, F., Mifsud, M., & Avila, L. V. (2021). A framework for the implementation of the sustainable development goals in university programmes. *Journal of Cleaner Production, 299.* https://doi.org/10.1016/j.jclepro.2021.126915

Leal Filho, W., Levesque, V., Sivapalan, S., Lange Salvia, A., Fritzen, B., Deckert, R., Kozlova, V., LeVasseur, T. J., Emblen-Perry, K., Azeiteiro, U. M., Paço, A., Borsari, B., & Shiel, C. (2022). Social values and sustainable development: Community experiences. *Environmental Sciences Europe, 34*(67), 1–13.

Li, W., & Yi, P. (2020). Assessment of city sustainability-Coupling coordinated development among economy, society and environment. *Journal of Cleaner Production, 256.* https://doi.org/10.1016/j.jclepro.2020.120453

Lin, T., & Lei, J. (2021). English-medium instruction and content learning in higher education: Effects of medium of instruction, English proficiency, and academic ability. *Sage Open, 11*(4) https://journals.sagepub.com/doi/pdf/10.1177/21582440211061533

Little, J. P. (2005). *Une conquête morale: L'enseignement en A.O.F.* L'Harmattan.

Louanchi, T. (1969). L'Algérie expérimente l'enseignement programmé et Télévisé. *Audio-Visual Media, 3*(2), 21–27.

Lozano, R., Ceulemans, K., Alonso-Almeida, M., Huisingh, D., Lozano, F. J., Waas, T., Lambrechts, W., Lukman, R., & Hugé, J. (2015). A review of commitment and implementation of sustainable development in higher education: Results from a worldwide survey. *Journal of Cleaner Production, 108*, 1–18.

Lozano, R., & Watson, M. K. (2013). Chemistry education for sustainability: Assessing the chemistry curricula at Cardiff University. *Educación Química, 24*(2), 184–192.

Lüdeke-Freund, F., Carroux, S., Joyce, A., Massa, L., & Breuer, H. (2018). The sustainable business model pattern taxonomy – 45 patterns to support sustainability-oriented business model innovation. *Sustainable Production and Consumption, 15*, 145–162.

Macaro, E. (2018). *English medium instruction*. Oxford University Press.

Mackey, W. (1989). Determining the status and function of languages in multinational societies. In U. Ammon (Ed.), *Status and function of languages and language varieties* (pp. 3–20). De Gruyter.

Mahdjoub, R., & Miliani, M. (2017). Education and career guidance in Algeria: Recurrent dysfunctions. In R. G. Sultana (Ed.), *Career guidance and livelihood planning across the Mediterranean: Challenging transitions in South Europe and the MENA region* (pp. 123–138). Sense Publishers.

Manière, L. (2010). La politique française pour l'adaptation de l'enseignement en Afrique après les indépendances (1958–1964). *Histoire de L'education, 128*, 163–190.

Maraf, B., & Vanci Osam, U. (2022). The smile revolution (hirak) as a driving force for an English 'tidal wave' and foreign language policy-making in Algeria. *Current Issues in Language Planning, 24*(2), 179–200.

Marginson, S. (1999). After globalization: Emerging politics of education. *Journal of education Policy, 14*(1), 19–31.

Marginson, S., & Rhoades, G. (2002). Beyond national states, markets, and systems of higher education: A glonacal agency heuristic. *Higher Education, 43*, 281–309.

Martin, I. (2003). Algeria's political economy (1999–2002): An economic solution to the crisis? *The Journal of North African Studies, 8*(2), 34–74.

Martinez, L. (2004). Why the violence in Algeria? *The Journal of North African Studies, 9*(2), 14–27.

Matasci, D., Jerónimo, M. B., & Dores, H. G. (Eds.). (2020). *Education and development in colonial and postcolonial Africa: Policies, paradigms, and entanglements, 1890s–1980s*. Cham: Palgrave Macmillan.

Matsumoto, A. (2019). Literature review on education reform in the UAE. *International Journal of Educational Reform, 28*(1), 4–23.

Mauviel, M. (2014). Le syndicalisme enseignant après l'indépendance. In A. Kadri (Ed.), *Instituteurs et enseignants en Algérie 1945–1975: Histoire et mémoires* (pp. 171–174). Karthala.

McKeown, R., Hopkins, C. A., Rizi, R., & Chrystalbridge, M. (2002). *Education for sustainable development toolkit*. University of Tennessee.

MESRS. (2021). Almutamar althaamin eashar liwuzara' altaelim aleali walbahth al ilmii fi alwatan ale arabi: nizam altaelim aleali walbahth al ilmii fi aljazayir, alwade alraahin wa efaq almustaqbal (The eighteenth conference of Ministers of higher education and scientific research in the Arab World: higher education and scientific research in Algeria, the current situation and future prospects) [*conference report*], Algiers. https://alecso.org/cmae/algerie.pdf

MESRS. (2022). *Grands agrégats* [Large aggregates]. https://www.mesrs.dz/index.php/fr/agregats-2/

MESRS. (2022). *Le réseau universitaire* [University network]. https://www.mesrs.dz/index.php/fr/agregats-2/

MESRS. (2022). *Portail des plateformes numériques* [Digital platforms gate]. https://www.mesrs.dz/index.php/fr/plateformes-mesrs/

MESRS. (2023). *M. Baddari présent les grands axes du budget du secteur pour l'année 2023* [Mr. Baddari outlines the primary focus areas of the sector's budget for the year 2023]. https://www.mesrs.dz/index.php/fr/2022/11/m-baddari-present-les-grands-axes-du-budget-du-secteur-pour-lannee-2023/

MESRS. (2023). *Majmueat min alkararat alwizaria* [A set of ministerial decisions]. https://www.mesrs.dz/index.php/2023/01/arretes-ministeriels-ar/

Meuleman, L. (2013). *Transgovernance: Advancing sustainability governance*. Springer Nature.

Miliani, M. (2012). Teaching in higher education institutions in Algeria: A clash of paedagogies? *International Journal of Pedagogies and Learning, 7*(3), 218–226.

Ministère de l'Environnement et des Energies Renouvelables. (2022). *Changements climatiques* (Climatic changes). https://www.me.gov.dz/fr/changements-climatiques/

Ministère de l'Environnement et des Energies Renouvelables. (2022). *Stratégie nationale de gestion intégrée et de valorisation des déchets à l'horizon 2035* [National strategy for integrated management and recovery of waste by 2035]. https://www.me.gov.dz/fr/environnement-urbain/dechets/

Moore, J. (2005). Is higher education ready for transformative learning? A question explored in the study of sustainability. *Journal of Transformative Education, 3*(1), 76–91.

Mostari, H. A. (2004). A sociolinguistic perspective on Arabisation and language use in Algeria. *Language Problems and Language Planning, 28*(1), 25–43.

Musso, M. (2020). Oil companies as agents of post-colonial relations: France, Algeria, and Italy in the Sahara. In V. Dimier & S. Stockwell (Eds.), *The business of development in post-colonial Africa* (pp. 127–156). Palgrave Macmillan.

Naylor, P. C. (2000). *France and Algeria: A history of decolonization and transformation*. University Press of Florida.

Northey, J. A., & Guemar, L. N. (2020). The Algerian hirak: Youth mobilisation, elections and prospects for reform. *Economics and Culture of the Middle East, 2020*(II), 14–21.

Nurunnabi, M. (2017). Transformation from an oil-based economy to a knowledge-based economy in Saudi Arabia: The direction of Saudi Vision 2030. *Journal of the Knowledge Economy, 8*(2), 536–564.

Nuseir, M. T., & El Refae, G. A. (2021). Factors influencing the choice of studying at UAE universities: An empirical research on the adoption of educational marketing strategies. *Journal of Marketing for Higher Education, 32*(2), 215–237.

O'Neill, M., & Chapman, A. (2015). Globalisation, internationalisation and English language: Studies of education in Singapore, Malaysia and Australia. *Education Research and Perspectives, 42*, 1–24.

O'Sullivan, A. (2013). Globalized education policy flow – Taking "the best"? *Proceedings of the Second International Conference on Emerging Research Paradigms in Business and Social Sciences (ERPBSS) held in Dubai from 26–28 November 2013*. https://www.academia.edu/5381630/Globalized_Education_Policy_Flow_-_-_Taking_the_best_

O'Sullivan, K. (2015). Education reform in the UAE – Bringing private schools into the fold. *Journal of Teaching and Education, 4*(1), 311–320.

Omar, O., & El-Bastawissi, I. (2023). Bringing the SDGs to life by 2030: Best practices in higher education within the Middle East region. In W. R. Filho, I. R. Abubakar, I. D. Silva, R. Pretorius, & K. Tarabieh (Eds.), *SDGs in Africa and the Middle East region* (pp. 1–26). Springer.

Palumbo, E., & Verga, F. (2015). Creation of an integrated environment to supply e-learning platforms with Office Automation features. *Interactive Learning Environments, 23*(6), 766–777.

Parker, P. D., Martin, A. J., Colmar, S., & Liem, G. A. (2012). Teachers' workplace well-being: Exploring a process model of goal orientation, coping behavior, engagement, and burnout. *Teaching and Teacher Education, 28*(4), 503–513.

Patashnik, E. (2003). After the public interest prevails: The political sustainability of policy reform. *Governance, 16*(2), 203–234.

Phuong, Y. H., & Nguyen, T. T. (2019). Students' perceptions towards the benefits and drawbacks of EMI classes. *English Language Teaching, 12*(5), 88–100.

Piao, X., & Managi, S. (2023). The international role of education in sustainable lifestyles and economic development. *Scientific Reports, 13*, 1–12.

Pun, J., Thomas, N., & Bowen, N. E. J. A. (2022). Questioning the sustainability of English-medium instruction policy in science classrooms: Teachers' and students' experiences at a Hong Kong secondary school. *Sustainability, 14*(4), 2168.

Rachidi, I., & Aballagh, A. (2023). Algeria's Hirak: Fading prospects for democratisation? *The Journal of North African Studies, 28*(6), 1515–1537.

Räisänen, C., & Fortanet-Gómez, I. (2008). The state of ESP teaching and learning in Western European higher education after Bologna. In C. Räisänen & I. Fortanet-Gómez (Eds.), *ESP in European higher education: Integrating language and content* (pp. 11–51). John Benjamins B.V.

Revere, R. B. (1971). *Consensus in independent Algeria, 1962–1965. PhD thesis.* Doctoral thesis, New York University. ProQuest Dissertations and Theses Global. https://www.proquest.com/docview/302435680/B28BF0FEEDB34AF4PQ/1?accountid=202267&sourcetype=Dissertations%20&%20Theses

Rezig, N. (2011). Teaching English in Algeria and educational reforms: An overview on the factors entailing students failure in learning foreign languages at university. *Procedia – Social and Behavioral Sciences, 29*, 1327–1333.

Rieckmann, M. (2018). Learning to transform the world: Key competencies in education for sustainable development. In A. Leicht, J. Heiss, & W. J. Byun (Eds.), *Issues and trends in education for sustainable development* (pp. 39–59). UNESCO.

Rose, H., Curle, S., Aizawa, I., & Thompson, G. (2019). What drives success in English medium taught courses? The interplay between language proficiency, academic skills, and motivation. *Studies in Higher Education, 45*(11), 1–14.

Rouaba, A. (2022). Algerian leader in bold move to promote English at junior school. *BBC online article*. https://www.bbc.com/news/world-africa-62368931

Ruedy, J. (2005). *Modern Algeria: The origins and development of a nation* (2nd ed.). Indiana University Press.

Ruiz-Mallén, I., & Heras, M. (2020). What sustainability? Higher education institutions' pathways to reach the agenda 2030 goals. *Sustainability, 12*(4), 1–18.

Sadiki, L., & Saleh, L. (2023). Degeneration and the demos in North Africa: Towards a 'critical' study of democratisation? *The Journal of North African Studies, 28*(6), 1444–1472.

Salem, S. (2021). *Global citizenship education in English as a foreign language (EFL) classrooms: Views of the Algerian university oral expression teachers and their students.* Doctoral dissertation, University of York.

Saturi, A. (2016). Altanmia almustadama fi aljazayar: alwaqie waltahadiyat (Sustainable development in Algeria: Reality and challenges). *Majalat Albahith, 16*(16), 299–311.

Saulière, J. (2014). Corporate language: The blind spot of language policy? Reflections on France's Loi Toubon. *Current Issues in Language Planning, 15*(2), 220–235.

Sayaf, A. M. (2023). Adoption of e-learning systems: An integration of ISSM and constructivism theories in higher education. *Heliyon, 9*(2), 1–17.

Schileo, G., & Grancini, G. (2021). Lead or no lead? Availability, toxicity, sustainability and environmental impact of lead-free perovskite solar cells. *Journal of Materials Chemistry C, 9*(1), 67–76.

Schmidt-Crawford, D. A., Lindstrom, D. L., & Thompson, A. D. (2023). The ISTE pledge: Committing to digital equity and transformation. *Journal of Digital Learning in Teacher Education, 39*(1), 2–3.

Schneider, E. W. (2011). *English around the world: An introduction.* Cambridge University Press.

Schoelen, L. (2023). Between national identity, research and social function – Academics' perceptions of the ambivalent role of the Algerian university. In B. K. Daniel & R. Bisaso (Eds.), *Higher education in Sub-Saharan Africa in the 21st century* (pp. 221–233). Springer.

Scoffham, S., & Rawlinson, S. (2022). *Sustainability education: A classroom guide.* Bloomsbury.

Scoones, I. (2016). The politics of sustainability and development. *Annual Review of Environment and Resources, 41*, 293–319.

Sebaaly, M. (2019). Digital transformation and quality, efficiency, and flexibility in Arab universities. In A. Badran, E. Baydoun, & J. R. Hillman (Eds.), *Major challenges facing higher education in the Arab World: Quality assurance and relevance* (pp. 167–178). Springer.

Serafini, P. G., de Moura, J. M., de Almeida, M. R., & de Rezende, J. F. D. (2022). Sustainable development goals in higher education institutions: A systematic literature review. *Journal of Cleaner Production, 370.* https://doi.org/10.1016/j.jclepro.2022.133473

Sever, S. D., & Tok, M. E. (2023). Education for Sustainable Development in Qatar. In L. Cochrane & R. Al-Hababi (Eds.), *Sustainable Qatar: Social, political and environmental perspectives* (pp. 329–347). Springer.

Sharma, A., Iyer, G. R., Mehrotra, A., & Krishnan, R. (2008). Sustainability and business-to-business marketing: A framework and implications. *Industrial Marketing Management, 39*(2), 330–341.

Siegel, J. (2020). Comprehension in English Medium Instruction (EMI) lectures: On the impact of lecturer L2 English use. *Language Learning in Higher Education, 10*(1), 73–91.

Silverstein, P. A. (2003). Martyrs and patriots: Ethnic, national and transnational dimensions of Kabyle politics. In J. McDougall (Ed.), *Nation, society and culture in North Africa* (pp. 85–109). Frank Cass.

Smit, U. (2019). Classroom discourse in EMI: On the dynamics of multilingual practices. In K. Murata (Ed.), *English-medium instruction from an English as a lingua franca perspective* (pp. 99–122). Routledge.

Snow, M. A. (2001). Content-based and immersion models for second and foreign language teaching. In M. Celce-Murica (Ed.), *Teaching English as a second or foreign language* (3rd ed., pp. 303–318). Heinle & Heinle Thompson Learning.

Stake, R. E. (1978). The case study method in social inquiry. *Educational Researcher, 7*(2), 5–8.
Sterling, S. (2001). *Sustainable education: Re-visioning learning and change* (Schumacher Briefings No. 6). Green Books, Ltd.
Sustainable Development Report. (2023). *Implementing the SDG stimulus: Includes the SDG index and dashboards.* https://s3.amazonaws.com/sustainabledevelopment.report/2023/sustainable-development-report-2023.pdf
Sutton, R. E., & Wheatley, K. F. (2003). Teachers' emotions and teaching: A review of the literature and directions for future research. *Educational Psychology Review, 15*, 327–358.
Tabari, R. (2014). *Education reform in the UAE: An investigation of teachers' views of change and factors impeding reforms in Ras Al Khaimah schools.* Sheikh Saud bin Saqr Al Qasimi Foundation for Policy Research.
The Brundtland Report. (1987). Report of the World Commission on Environment and Development: Our Common Future. Retrieved from: https://sustainabledevelopment.un.org/content/documents/5987our-common-future.pdf
The New Arab. (2022). *Algeria recruits 5,000 new English language teachers for primary schools in shift from French.* https://www.newarab.com/news/algeria-hires-5000-english-teachers-shift-french
The World Bank. (2022). *GNI (current US$) – Algeria, Saudi Arabia.* https://data.worldbank.org/indicator/NY.GNP.MKTP.CD?locations=DZ-SA
Tillion, G. (1958). *Algeria: The realities* (R. Matthews, Trans.). Alfred A. Knopf.
Tilmatine, M. (2015). Arabisation and linguistic domination: Berber and Arabic in the North of Africa. In C. Stolz (Ed.), *Language empires in comparative perspective* (pp. 1–16). De Gruyter.
Tran, N. T., Tran, T. B. T., & Bien, T. T. M. (2020). An exploration of the factors hindering students' lesson comprehension in EMI classes. *International Journal of Language and Literary Studies, 2*(3), 29–42.
Trend, R., Davis, N., & Loveless, A. (1999). *QTS: Information and communications technology.* Letts.
UNESCO. (2016). *Education 2030: Incheon declaration and framework for action for the implementation of sustainable development goal 4.* UNESCO. https://unesdoc.unesco.org/ark:/48223/pf0000245656
UNESCO. (2018). *Paving the road to education: A target-by-target Analysis of SDG 4 for Asia and the Pacific.* https://unesdoc.unesco.org/ark:/48223/pf0000265912
UNESCO. (2023). *What you need to know about education for sustainable development.* https://www.unesco.org/en/education-sustainable-development/need-know
Uysal, H. H., Plakans, L., & Dembovskaya, S. (2008). English language spread in local contexts: Turkey, Latvia and France. *Current Issues in Language Planning, 8*(2), 192–207.
Van der Walt, C. (2013). *Multilingual higher education: Beyond English medium orientations* (Vol. 91). Multilingual Matters.
Vu, N. T., & Burns, A. (2014). English as a medium of instruction: Challenges for Vietnamese tertiary lecturers. *Journal of Asia TEFL, 11*(3), 1–31.
Wade, R. (2008). Education for sustainability: Challenges and opportunities. In J. Coriddi (Ed.), *Policy & practice: A development education review* (pp. 30–48). NI Ltd.
Walton, J. (2000). Should monitoring be compulsory within voluntary environmental agreements? *Sustainable Development, 8*(3), 146–154.

Webb, A., & Rodriguez, L. (2022). Sustainable development goals and universities: Higher education engagement through the global master's in development practice programme. In M. Öztürk (Ed.), *Engagement with sustainable development in higher education: Universities as transformative spaces for sustainable futures* (pp. 1–18). Springer.

Weber, J. J., & Horner, K. (2012). *Introducing multilingualism: A social approach.* Routledge.

Weber, L. (2014). Louis Weber, entre syndicalisme et politique. In A. Kadri (Ed.), *Instituteurs et enseignants en Algérie 1945–1975: Histoire et mémoires* (pp. 249–260). Karthala.

Werenfels, I. (2007). *Managing instability in Algeria: Elites and political change since 1995.* Routledge.

Wilson, C. (2013). *Credible checklists and quality questionnaires: A user-centered design method.* Elsevier.

Yilmaz, T., Li, S., & Lee, Y. J. (2024). A multilingualism-as-resource approach in a graduate-level English-medium instruction (EMI) program at a Korean university. *Journal of Multilingual and Multicultural Development, 45*(1), 1–14.

Yin, R. K. (2003). *Case study research: Design and methods* (3rd ed.). SAGE Publications.

Zalazar-Jaime, M. F., Moretti, L. S., García-Batista, Z. E., & Medrano, L. A. (2023). Evaluation of an academic satisfaction model in e-learning education contexts. *Interactive Learning Environments, 31*(7), 4687–4697.

Zartman, I. W. (1975). Algeria: A post-revolutionary elite. In F. Tachau (Ed.), *Political elites and political development in the Middle East: States and societies of the Third World* (pp. 255–292). Schenkman Publishing Company.

Zeraoulia, F. (2020). The memory of the Civil War in Algeria: Lessons from the past with reference to the Algerian Hirak. *Contemporary Review of the Middle East, 7*(1), 25–53.

Zeraoulia, F. (2021). National reconciliation in Algeria from a bottom-up approach: Analysing victims' narratives. *The Journal of North African Studies, 7*(5), 862–893.

Zhou, Y., Zhao, J., & Zhang, J. (2020). Prediction of learners' dropout in e-learning based on the unusual behaviors. *Interactive Learning Environments, 31*(3), 1796–1820.

Index[1]

[1] Note: Page numbers followed by 'n' refer to notes.

S. Bouherar, S. Salem, *Sustainability in Algerian Higher Education*, https://doi.org/10.1007/978-3-031-81347-4

J

L

M

N

O

The manufacturer's authorised representative in the EU is Springer Nature Customer Service Centre GmbH, Europaplatz 3, 69115 Heidelberg, Germany. If you have any concerns regarding our products, please contact ProductSafety@springernature.com

Printed and bound by CPI Group (UK) Ltd, Croydon, CR0 4YY
01/10/2026
02225953-0017